Sleepers, *Wake*!

Jacket: 'Atom Piece', 1963–4, by Henry Moore

BARRY JONES

Sleepers, *Wake*!

Technology and the future of work

Melbourne
OXFORD UNIVERSITY PRESS
Oxford Auckland New York

Oxford University Press

LONDON GLASGOW NEW YORK TORONTO
DELHI BOMBAY CALCUTTA MADRAS KARACHI
KUALA LUMPUR SINGAPORE HONG KONG TOKYO
NAIROBI DAR ES SALAAM CAPE TOWN
MELBOURNE AUCKLAND
and associates in
BEIRUT BERLIN IBADAN MEXICO CITY NICOSIA

First published 1982
Reprinted 1982

Jones, Barry, 1932-
Sleepers, Wake!

Bibliography.
Includes index.
ISBN 0 19 554270 3 (paperback)
ISBN 0 19 554343 2 (casebound)

1. Social change. 2. Technology – Social aspects. 3. Technocracy. 4. Technologists. 5. Labor supply – Australia. 6. Unemployment, Technological – Australia. I. Title.

303.4'83

Designed by Arthur Stokes

Computer photocomposed at
Griffin Press Limited, Netley, South Australia.
Printed by Richard Clay (S.E.Asia) Pte Ltd

PUBLISHED BY OXFORD UNIVERSITY PRESS, 7 BOWEN CRESCENT, MELBOURNE

PRINTED IN SINGAPORE

To Lewis Mumford
Simon Nora
Hugh Stretton

Contents

Preface

Sleepers, wake! The watchman on the heights is calling . . .
J. S. Bach (Cantata no. 140, *Wachet auf*)

Throughout the 1970s, the technologically advanced world underwent a post-industrial revolution which will change society in the 1980s more rapidly than any decade since the 1780s when the Industrial Revolution began in Great Britain. In the 1970s, politicians, bureaucrats and economists were too close to post-industrial changes to see them in perspective, recognize what was happening and work out appropriate responses. A downturn in economic growth rates and sharp rises in unemployment led to a collapse of confidence in the political process and a turning towards simplistic solutions for complex economic and social problems.

In the 1980s, new techniques can decimate the labour force in the goods-producing sectors of the economy. This will either perpetuate massive unemployment or lead to the creation of large-scale, low-output 'servile' work in the service sector. There will be a vast increase in transactions based on the collection, manipulation and dissemination of information by computerized technologies. This will bring about a fundamental change in the relationship of people and work. Lewis Mumford condensed this into a brief formula:

> Manual work into machine work: machine work into paper work: paper work into electronic simulation of work, divorced progressively from any organic functions or human purposes, except those that further the power system.

This process may destroy the fragile consensus on which the democratic system depends. Technology is a political instrument and becomes an end in itself. Power will move towards the controllers of technology and away from a poorly informed and increasingly apathetic electorate. The democratic system, increasingly irrelevant in the 1970s, may become obsolete in the 1980s. Of the ten or twelve basic tasks carried out by members of parliament, every one could be carried out more cheaply and effectively by computer. Who needs us if utilitarianism, the ideology of the machine, is strictly applied? What impact will the loss of faculty have on the human species? If *homo sapiens* is no longer *homo faber*, if he ceases to be *homo laborans* or even *homo sedentarius* and is transformed to *homo ludens*, what is the future of the race?

The problems raised in this book transcend the political ideologies of left and right: the issues are as applicable in the USSR as in the USA, or in China and Mexico. They are equally relevant to the Third World, where the greatest population movement of all time is going on now – from the country to the city.

Recent research on the operation of the human brain suggests a profound dichotomy between the left and right hemispheres. The left hemisphere, it is argued, dictates behaviour that is analytical, reductionist, rule-following, verbal, aggressive, competitive and linear, conforming to the masculine stereotype. The right hemisphere, more closely conforming to the feminine stereotype, is related to perceptions which are holistic, transcendental, impressionistic, visual, co-operative and lateral – with strong emphasis on seeing things in a broad context and linking them together.

Temperamentally, I am an optimist. Just as Francis Crick applied lateral thinking to solve the problem of the structure of DNA, which had defied orthodox analysis and helped to create molecular biology and biotechnology (areas in which he had no formal training), so I believe that new approaches to social and economic problems can make the 1980s a creative era in which Mozartian man (and woman), as Dennis Gabor wrote, can evolve. If we remain imprisoned in the linear thinking so congenial to bureaucrats, capitalists, commissars and aspiring gauleiters, the 1980s will be a period of unemployment, alienation and unprecedented social crises.

Barry Jones
Melbourne
1981

Acknowledgments

Acknowledgment for permission to reproduce material from the following books is gratefully made to: Macmillan Publishers Ltd, London and Basingstoke: J. M. Keynes, *Essays in Persuasion*, 1931. Macmillan Publishers Ltd, London and Basingstoke and Peters and Co., London and Harper and Row Inc.: David McLellan, *Marx's Grundrisse*, 1971. Prentice-Hall Inc.: James Martin and Adrian Norman, *The Computerized Society*, 1973. The US Department of Commerce: M. Porat, *The Information Economy*, vol. 1, 1977. Cambridge University Press: H. Stretton, *Capitalism, Socialism and the Environment*, 1976. Progress Publishers, Moscow: V. I. Lenin, *Collected Works*, vol. 3, 1974; Karl Marx, *Capital*, 1974. George Allen & Unwin (Publishers) Ltd: Bertrand Russell, *In Praise of Idleness and Other Essays*, 1976. Martin Secker & Warburg Limited and Alfred A. Knopf Inc.: Dennis Gabor, *Inventing the Future*, 1964. Marion Boyars Ltd and Pantheon Books Inc.: Ivan Illich, *Energy and Equity*, 1973. Perigee Books: Hazel Henderson, *Creating Alternative Futures*, 1975 © 1975, Ron Bernstein Agency. MIT Press: Simon Nora and Alain Minc, *The Computerisation of Society*, 1980; Langdon Winner, *Autonomous Technology*, 1979. Foundation for Australian Resources: B. S. Thornton and P. M. Stanley, *Computers in Australia: Usage and Effects*, 1978. Penguin Books Australia Ltd: Peter Sheehan, *Crisis In Abundance*, 1980. Academic Press Inc. and Professor Randall Collins: R. Collins, *The Credential Society: An Historical Sociology of Education and Stratification*, 1980. Anderla, OECD: Georges Anderla, *Information in 1985*, 1973. Granada Publishing Limited: E. J. Mishan, *The Costs of Economic Growth*, 1968. North Holland Press and North Holland Publishing: Y. Masuda, 'Future Perspectives for Information Utility' in *Evolution in Computer Communications*,

1978. Weidenfeld & Nicolson: Albert Speer, *Inside the Third Reich*, 1970. Martin Secker & Warburg Limited and Harcourt Brace Jovanovich, Inc.: Lewis Mumford, *The Pentagon of Power*, 1967. Alfred A. Knopf, Inc. and Jonathon Cape Ltd: J. Ellul, *The Technological Society*, trans. J. Wilkinson, 1964. Fontana Books (Collins): D. Dickson, *Alternative Technology*, Glasgow, 1974. By permission of Penguin Books Ltd an extract is reprinted from Karl Marx, *Capital*, vol. 1, trans. Ben Fowkes (The Pelican Marx Library/New Left Review, 1976), edition and notes copyright © New Left Review, 1976; translation copyright © Ben Fowkes, 1976. By permission of Penguin Books Ltd an extract is reprinted from J. A. Trevithick, *Inflation* (Pelican Books, 1977), p. 56, copyright © J. A. Trevithick, 1977. An extract is reprinted from Joseph Weizenbaum, *Computer Power and Reason: From Judgement to Calculation*, W. H. Freeman and Company, copyright © 1976. An extract is reprinted from Octavio Paz, 'Mexico and the United States', *New Yorker*, 17 September 1979, reprinted by permission © 1979 The New Yorker Magazine, Inc. A cartoon by Levin 'Work-O-Mat' is reprinted from *New Yorker*, 19 November 1979; © 1979 The New Yorker Magazine, Inc. An extract from H. Braverman, *Labour and Monopoly Capital: the Degradation of Work in the Twentieth Century* (1976 © 1974 by Harry Braverman) is reprinted by permission of Monthly Review Press.

My thanks to Henry Moore for his permission to use the photograph of 'Atom Piece' on the cover of this book.

I am grateful for the assistance and advice of many people including Clyde Cameron, Frances Cushing, Sarah Dawson, Jonathan Gershuny, Frank Hainsworth, Rosemary Hanbury, Patricia Kennedy, Mary Lloyd, Catherine McDonald, Peter McGregor, Ieuan Maddock, Kaye Manning, Gerry Newman, Simon Nora, Emma Rothschild, Garth Simmonds, Hugh Stretton, Louise Sweetland, Vivian Wilson, Shirley Williams, Langdon Winner, Michael Young, and Mick Young.

Disclaimer

1 From a Pre-Industrial to a Post-Service Society

The microprocessor has finally repealed the labour theory of value; there is really no possibility now of maintaining the fiction that human beings can be paid in terms of their labour. The link between jobs and income has been broken . . .

Hazel Henderson

The Industrial Revolution began in Great Britain in the 1780s,[1] spread through Europe and North America in the nineteenth century, and in the twentieth century became the preferred model for worldwide economic development. After the Industrial Revolution, for the first time in human history, agriculture ceased to employ a majority of workers and began its long and continuous decline as an employer – but not as a producer. During the industrial era, manufacturing and directly related industrial employment (transport, retailing, mining) dominated the labour market. Towns and cities grew rapidly, self-sufficiency declined and specialization (the division of labour) intensified. Industrial employment reached its highest proportion of the labour force in Australia after 1947, the United States in 1950, Great Britain in 1951, New Zealand in 1956, Belgium in 1963, France and Canada in 1964, Sweden in 1965, West Germany in 1968, and Japan in 1970. In each of these countries, industry then began to decline as an employment sector – but not as a producer. (Australia was something of a special case: industrial employment reached a plateau in 1947 and stayed there, with insignificant fluctuations, until 1965 when it began its decline.)

Advanced economies have now moved into a post-industrial era, in which services such as welfare, education, administration and the transfer of information dominate employment. The displacement of agriculture by manufacturing as the dominant employer was the first of two major 'cross-overs' in economic history. The second was the displacement of manufacturing (industrial) employment by service (post-industrial) employment. The post-industrial era will be of short duration: the technological revolution of the 1980s will bring about a third major transition – to a 'post-service' society in which routine and repetitive service employment will be significantly reduced, or eliminated. This change will raise unprecedented human problems: the

Towards a post-service economy

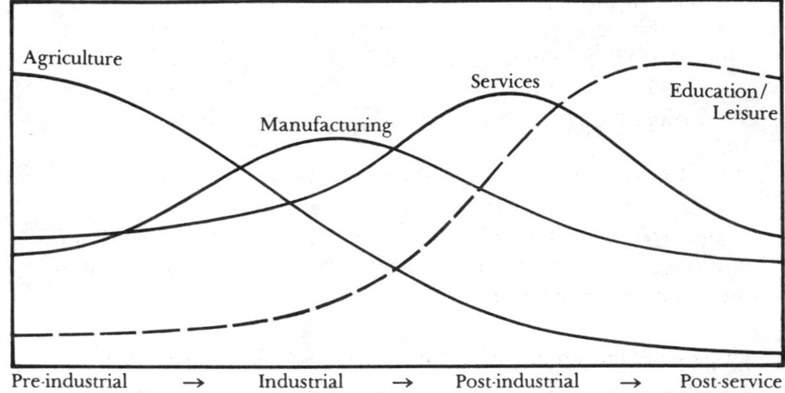

whole relationship of people to time use, personal goals, economics, politics and culture must be re-examined.

The stages of development

PRE-INDUSTRIAL TO INDUSTRIAL

The term 'Industrial Revolution' was adopted surprisingly late in English. Auguste Comte's periodical *L'Industrie* coined the word *industrialisme* (1816), while Auguste Blanqui described *la révolution industrielle* in his *Histoire de l'économie politique* (1837). Friedrich Engels (1845) and Karl Marx (1863) both used the term – but its first English usage was in *Lectures on the Industrial Revolution of the Eighteenth Century in England* (1884), a posthumous work by Arnold Toynbee. The Industrial Revolution was marked by the following elements:

1 Small-scale subsistence farming declined and was replaced by large-scale, high-volume, commercial agriculture. Farm workers fell below 50 per cent of the labour force in Britain before 1780, Belgium by 1850, the United States (eastern states) before 1860, Australia before 1870, Germany in 1875, France about 1890 and the USSR in 1947.

2 'Industry' (i.e. manufacturing and mining) passed agriculture as an employer in Britain about 1815, Belgium before 1890, Germany in 1896, Australia in 1900, the United States in 1907, France in 1913, New Zealand in 1940, and the USSR in 1960.

3 Employment based at or near home declined, and was replaced by work at central locations such as factories and shops. This led to the new concept of 'going to work' and the rise of a new class, the indus-

trial proletariat. This in turn led to the development of 'industrial discipline' and the inevitable evolution of trade unions.

4 Urban population grew very rapidly (and in conditions of great squalor in the first decades), reaching 50 per cent of the total in Britain in 1851, Belgium in the 1850s, Australia by 1870, Germany by 1895, the United States about 1910, France in 1931 and the USSR about 1950.

5 Intensification of the division of labour led to the elimination of self-employed craftsmen, the de-skilling and elimination of ancient trades, the introduction of unskilled 'process work' and machine-minding, often performed by women and children and leading to a reduction in wage levels.

Employment trends in Great Britain since 1780, by three sectors

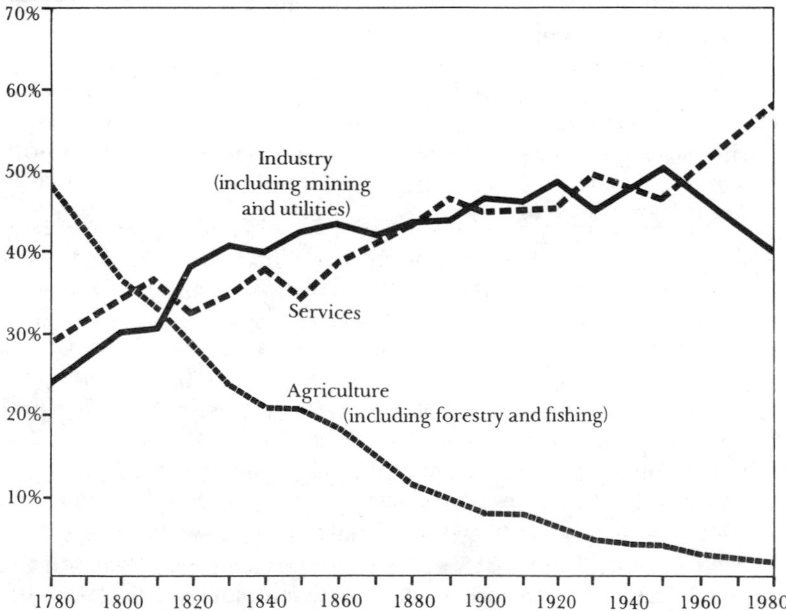

Note: The figures for the earlier years are approximate, following Cipolla, Hartwell and Kuznets, to illustrate trends. These trends appeared *first* in Britain, and similar graphs – with later starting dates – could be drawn for Belgium (from 1825), the United States (1830–40), France and Australia (1850–70). The lines indicate general trends only. The figures after 1801 are based on the analysis by Deane and Cole of census returns and refer only to persons actually employed (the unemployed are not counted and the huge dislocations caused during World Wars I and II are disregarded).

6 Machines replaced, or extended, human capacity. Animals were gradually eliminated as an energy source.

7 Industrial activity and production depended on standardization, co-ordination and synchronization: workers had to respond to the precision, rapidity and regularity of machines; factories and railways ran according to the clock; parts were made interchangeable.

8 The development of railways, steamboats, telegraphs, telephones and other forms of communication made possible the rapid dispersal of population – not only within cities but also throughout nations and overseas.

9 There was a vast increase in the use of iron, coal, copper and, later, oil and rubber, and a decline in the use of wood or animal substances. Demands for raw materials became central elements in foreign and colonial policies.

10 The development of mercantile, materialist and competitive values had a profound effect on politics, culture, science, philosophy and religion.

11 The rise of the urban middle class and increasing diversity in city life led to substantial increases in 'service' employment – servants, shopkeepers, railway and other transport workers. (In the 1891 British census, agriculture and domestic service were still the two largest occupational headings listed.)

INDUSTRIAL TO POST-INDUSTRIAL

The term 'post-industrial' was coined in 1913 by the art historian Ananda K. Coomaraswamy (1877–1947), and was taken up by the guild socialist Arthur Joseph Penty (1876–1937). It is used by those who reject excessive dependence on high technology, and also by enthusiasts for 'hyper-industrialism'. I use the term as a neutral description of an economy where the majority of the labour force is no longer employed in agriculture, mining, manufacturing, construction and directly related industrial services such as retailing, storage and transport. The post-industrial revolution has the following characteristics:

1 There is a historic transition in employment, marked by a decline in labour/time-saving (or capital-intensive) employment, which is intended to produce maximum outputs in value added with the smallest possible investment in labour, and a rise in service employment.

2 Employment in services – selling, teaching, office work, transport, administration, media and other labour/time-absorbing (labour-intensive) jobs – reached 50 per cent of the paid labour force in Australia about 1945, the United States by 1947, Great Britain about

1948, Canada by 1955, Sweden before 1960, New Zealand in 1965, Belgium in 1966, France and West Germany by 1972, and Japan by 1974. It is still rising: by 1980, over two-thirds of employment in sophisticated economies was in services, depending on definition.[2]

3 The economy is increasingly based on information, and more people are engaged in collecting, processing, storing and retrieving data than are employed in agriculture and manufacturing.

4 A global economy has been created, with a growth in power by multi-national corporations and a loss of power by individual nations which lack the capacity or will to control their own economic destinies. With instant worldwide communication and the development of an international division of labour, specific local or regional market forces are of decreasing relevance in determining patterns of demand and employment.

5 There is a vast increase in the volume of accessible information – and as the time available to individuals to read, consider and understand does not expand correspondingly, knowledge is becoming fragmented. Subjects dealt with by governments (such as budgets) are so complex that the decision-makers cannot consider and understand all the relevant factors: politics loses its ideological force, the electoral process has decreasing significance, and power moves away from elected governments towards interest groups dominating particular sectors of society – technocrats, public servants, corporations and media owners.

6 The post-war era of full employment (1945–74) has gone, and prevailing economic theories are vehemently opposed to government intervention to restore it. The demand for unskilled and semi-skilled workers is declining.

7 There is a widening gap between the 'information-rich' and the 'information-poor'. Educational qualifications have become rationing devices for entry into secure and satisfying employment, and failure imposes heavy social and economic penalties.

8 'Miniaturization', a product of micro-electronics, has made possible an exponential increase in output with a far lower investment in capital, labour and energy, marking a radical discontinuity with past economic history.

9 Cities making the post-industrial transition are likely to fall in population (e.g. New York and London as shown by the 1980 and 1981 censuses).

POST-INDUSTRIAL TO POST-SERVICE

Just as agriculture declined as a major employer in the nineteenth century to enable expansion in manufacturing to occur, and as manufacturing declined as a major employer in the twentieth century to enable expansion in services to occur, it now seems inevitable that market-based service employment will decline even more rapidly – in exactly the same way, and for exactly the same reasons – due to the introduction of miniaturized, sophisticated, low-cost technology. We seem likely to pass through a post-service revolution into a post-service society – which could be a golden age of leisure and personal development based on the co-operative use of resources. (The term 'post-service society' was coined by Bertram M. Gross in an article 'Planning in an Era of Social Revolution', published in 1971.[3]) But if we do not choose this option, and if things are allowed to drift, economic power will become even more unequally divided than it is now, resources will be the subject of a bitter struggle between the strong and the weak, and prolonged, massive unemployment will traumatize and divide society. The post-service society will have these features:

1 There will be a sharp reduction in the numbers of people working in large-scale, market-based service employment (e.g. office work, retailing, fuel supply, etc.) which is routine or repetitive and can be replaced by computerized operations which are cheaper, faster and more accurate.

2 Employment will also fall where personalized services can be provided by centralized, computer-controlled means (such as electronic mail delivery, telecommunications, printing) – offset to some extent by the growth of computer programming.

3 Manufacturing will use fewer workers, especially in small firms which are forced to computerize to cut labour costs and survive against international or domestic capital-intensive competition. With increasing entry to the global economy, employment in textiles, footwear, clothing, motor vehicles and accessories will fall. There may be some compensating increases in making solar-energy equipment and leisure goods, and processing and fabricating metals.

4 There will be a tendency towards a 'non-political' corporate state in which elected parliaments confine themselves to the contracting number of subjects which lay people feel competent to deal with. All political groupings may reach a consensus about 'technological determinism', whereby all complex problems are seen as essentially technical in nature and left to the technicians to resolve.

5 There will be increasing anxiety about the rate of depletion of the world's resources, and the adoption of a 'stable state' economy in which materials are recycled and miniaturized technology is used to save energy.

6 The traditional work ethic will be declared irrelevant or

counter-productive to society's needs. Compulsory leisure activities may be imposed on those for whom there is no place in the labour force. There should be a fundamental change in the relationship of workers to their employment, to time use, and to each other.

7 There will be mounting tension between the information-rich, who are employed and affluent, and the information-poor, who are unskilled, bored, frustrated and unhappy about subsisting on guaranteed income.

or

8 New types of employment will be generated which are *complementary to* and *not dependent on* technological forms. They will be aimed at satisfying individual needs – deliberately intended to be labour/time-absorbing – and work itself will be part of the output of production, as in craftwork, gardening, research, sport, leisure, hobby and do-it-yourself (DIY) activities.

We face the prospect which John Maynard Keynes wrote about in his essay 'Economic Possibilities for Our Grandchildren' in 1930. Keynes (1883–1946) was a British economist whose book *The General Theory of Employment, Interest and Money* (1936) revolutionized government economic policies during and after World War II.

We are being afflicted with a new disease of which some readers may not yet have heard the name, but of which they will hear a great deal in the years to come – namely, *technological unemployment.* This means unemployment due to our discovery of means of economising the use of labour outrunning the pace at which we can find new uses for labour.

But this is only a temporary phase of maladjustment. All this means in the long run that *mankind is solving its economic problem* . . .

I draw the conclusion that, assuming no important wars and no important increase in population, the *economic problem* may be solved, or be at least within sight of solution, within a hundred years. This means that the economic problem is not – if we look into the future – *the permanent problem of the human race.*

Why, you may ask, is this so startling? It is startling because – if, instead of looking into the future, we look into the past – we find that the economic problem, the struggle for subsistence, always has been hitherto the primary, most pressing problem of the human race – not only of the human race, but of the whole of the biological kingdom from the beginnings of life in its most primitive forms.

Thus we have been expressly evolved by nature – with all our impulses and deepest instincts – for the purpose of solving the economic problem. If the economic problem is solved, mankind will be deprived of its traditional purpose.

Will this be a benefit? If one believes at all in the real values of life, the prospect at least opens up the possibility of benefit. Yet I think with dread of the readjustment of the habits and instincts of the ordinary man, bred into him for countless generations, which he may be asked to discard within a few decades . . .

Thus for the first time since his creation man will be faced with his real, his permanent problem – how to use his freedom from pressing economic cares, how to occupy the leisure, which science and compound interest will have won for him, to live wisely and agreeably and well.

The strenuous purposeful money-makers may carry all of us along with them into the lap of economic abundance. But it will be those peoples, who can keep alive, and cultivate into a fuller perfection, the art of life itself and do not sell themselves for the means of life, who will be able to enjoy the abundance when it comes.

Yet there is no country and no people, I think, who can look forward to the age of leisure and of abundance without a dread. It is a fearful problem for the ordinary person, with no special talents, to occupy himself, especially if he no longer has roots in the soil or in custom or in the beloved conventions of a traditional society . . .

For many ages to come the old Adam will be so strong in us that everybody will need to do *some* work if he is to be contented. We shall do more things for ourselves than is usual with the rich today, only too glad to have small duties and tasks and routines. But beyond this, we shall endeavour to spread the bread thin on the butter – to make what work there is still to be done to be as widely shared as possible. Three-hour shifts or a fifteen-hour week may put off the problem for a great while . . .

When the accumulation of wealth is no longer of high social importance, there will be great changes in the code of morals. We shall be able to rid ourselves of many of the pseudo-moral principles which have hag-ridden us for two hundred years, by which we have exalted some of the most distasteful of human qualities into the position of the highest virtues. We shall be able to afford to dare to assess the money-motive at its true value.[4]

The plan of the book

The foregoing pages set out the central argument of this book – that technologically based transitions create revolutionary economic and social changes. The thesis is controversial, disputed by many and arousing hostility if not contempt from most professional economists. With this in mind I have endeavoured to set out the basis of my theme as extensively as possible, both establishing the historical background and taking account of the wide variety of misgivings the argument may produce.

Chapter 2, 'An Age of Discontinuity', contends that Western civilization has passed through three interrelated Industrial Revolutions in two hundred years: the ages of steam (1780–1840), electricity (1860–1910) and atomic power (1942–70). The 1970s, it is argued, marks the beginning of a radically new post-industrial era in which labour-displacing technology will change the nature and quantity of work. Is this a break with the past? What are the arguments for and against the orthodox economists' belief in continuity? Will there be a general collapse of work? Have we reached an 'age of asymptote'?

Chapter 3, 'A New Analysis of the Labour Force', proposes that use of the conventional three-sector analysis be abandoned as it is no longer adequate to describe what people are actually doing. It explains the development of information-based employment, and urges that both this and home-based employment should be recognized as separate labour-force sectors. This five-sector analysis is then used to describe the Australian labour force historically, nationally and regionally, in order to identify likely areas of employment growth and decline and to demonstrate the class and regional basis of structural unemployment.

Chapter 4, 'Two Types of Employment and Time Use', distinguishes between labour/time-saving work (based on economies of scale) and labour/time-absorbing work (largely Parkinsonian), and considers the implications for labour-force planning. This chapter also examines the contrasting aims of the 'market sector' (based on capital and the profit motive) and the 'convivial sector' (aimed at meeting the community's non-economic needs), but concludes that in reality they are interdependent.

Chapter 5, 'Computers and Employment', describes computer-based technology, its range, capacity and cost, and the revolutionary impact of miniaturization on economics and labour requirements.

Chapter 6, 'Unemployment, Inflation, Demand and Productivity', examines Australia as it entered the 1980s and questions the assumptions on which various policies aimed at reducing unemployment are based:

1 'Beat inflation first' (Friedman).
2 'Stimulate demand' (Keynes).
3 'Increase productivity' (Okun).

Examination of disaggregated employment statistics suggests that employment levels in particular areas are closely related to three factors: access to higher education, the diversity of demand for goods and services, and the complexity of local infrastructures.

Chapter 7, 'Education and Employment', describes the wide gap between the expectations and participation rates of the children of middle-class and working-class families, and the effect this has on the increasing competition for jobs.

Chapter 8, 'The Information Explosion and Its Threats', argues that the information explosion places the democratic system under massive stress unless democratic, pluralist access to information is guaranteed.

Chapter 9, 'Work in an Age of Automata', discusses the survival of the traditional work ethic, its growing irrelevance in an age of automata, and potential difficulties in finding future work for the unskilled and semi-skilled.

Chapter 10, 'Technological Determinism: "But We Have No

Choice" ', examines the doctrine of technological determinism and the conscious way in which particular forms of technology are pushed as if there were no alternatives. This has largely paralysed the political process in Australia.

Chapter 11, 'What Is To Be Done?', proposes a programme of political action to secure the best social consequences of rapid technological change. It also points out that it is even more important – and far more difficult – to raise levels of consciousness and encourage the development of active human responses. What is the political result of concentrating too much technological power in too few hands?

I have used Australian examples to illustrate my argument because Australia is the country I know best and the evidence is at hand, but the thesis is generally applicable to all technologically advanced societies.

2 An Age of Discontinuity

No, Sire, it is a revolution.

Duc de La Rochefoucauld-Liancourt to Louis XVI, 1789

The two most significant turning points in human history so far have been the Neolithic Agricultural Revolution and the Industrial Revolution. The Neolithic Revolution marked the transition from nomadic pastoral life, where man was dependent on hunting and food-gathering, to life in settlements based on systematic food-production.[1] It began in Western Asia in about 8000 BC, spread slowly through Europe and reached the British Isles about five thousand years later. Its achievements included the building of houses and towns, the use of axes, hammers, wheels and boats, spinning and weaving, mining and metal-working, making beer and bread, and the use of words, symbols and probably numbers. The Industrial Revolution swept through the world in less than two hundred years, reducing agricultural employment to a small fraction of the total labour force in industrialized countries in Europe, North America and Oceania. It was both a cause and an effect of an explosive increase in world population from 800 million in 1780 to 4500 million in 1980, with an enormous growth in cities. Manufacturing and the search for raw materials shaped the economic, political and moral development of the human race, leading to the apotheosis of capitalism, the rise of socialism and the decline of religion.

We are now entering an age of discontinuity, marking an end to the industrial era, in which the whole range of human capacity and experience may be changed beyond recognition in a few decades. I believe that the post-industrial era will be of short duration, and then transformed into a post-service society in which work as we know it – the dominant feature of human life outside the home throughout history – will be dramatically reduced. This will raise fundamental questions about the human condition which few seem able or willing to discuss. Before turning to the basic question of historical continuity or discontinuity, we must examine the nature of the Industrial Revolution.

Three industrial revolutions

The Industrial Revolution is often described as if it was a single entity, a steady development producing greater growth and wealth. A closer examination of the history of technological change suggests that there have in fact been three distinct but interrelated industrial revolutions, each marked by clusters of inventions and discoveries.[2]

The Steam Revolution. The First Industrial Revolution began in Britain and ran from about 1780 to 1840. Its greatest achievements were the application of steam power to textiles, mining, manufacturing and transport, in precision engineering, the beginnings of chemical synthesis, and experiments in the applied use of electricity. Many of the inventors were amateurs – Arkwright was a wig-maker, Cartwright a clergyman – who responded to deficiencies in already existing industries. Where one man had operated a single spindle in 1770, by 1840 his grandchild could operate one hundred. The inventions mostly responded to need: few new industrial forms were created.

The Electric Revolution. The Second Industrial Revolution occurred between 1860 and 1910, largely in the United States, Germany and Britain. It was based on applied technology, marked by huge increases in the production of steel and chemicals, the use of oil and electricity as an energy source for industry and transport, the electrification of cities, the telephone and telegraph, cheap photography, scientific medicine, the first plastics, motor cars, aeroplanes and radios. Inventions were often made by professionals working in related areas: Alexander Graham Bell was a teacher of the deaf who devised his telephone as a therapeutic aid; W. H. Perkin was trying to synthesize explosives when he produced the first aniline dye; Thomas Alva Edison discovered the operating principle of the electron tube but did not grasp its significance. In many cases need responded to the invention – there is little evidence of prior demand for the telephone, but after 1876 its very existence stimulated demand and created much employment for exchange operators. The decades after 1910 saw the refinement of inventions and their mass production, but few major innovations emerged.

Great Britain, having dominated the First Industrial Revolution, played a reduced role in the Second (and Third) due to rapid development in other countries. In 1850, it had 40 per cent of the world's manufacturing output, falling to 20 per cent in 1900, 14 per cent in 1914, and 4 per cent by 1963.

The Atomic Revolution. The Third Industrial Revolution began in the

United States in 1942 when Enrico Fermi began the first sustaining nuclear chain reaction at the University of Chicago. It is still in progress, although in decline since about 1970. Its sustaining theme has been the development of pure and applied research towards the exploration of extremities, and its major achievements include atomic fission and fusion to produce explosive quantities of energy, supersonic aircraft and missiles, radio astronomy, the investigation of outer space, planets and the moon, explaining the complexities of DNA and protein structure, computers, micro-electronics and the universal diffusion of synthetic materials, television, civil aviation and the motor car. Most discoveries have resulted from large-scale collective research and development, generally under government sponsorship, and were not aimed specifically at producing new goods and services. After 1970 there were many refinements of already existing technologies, but it is hard to identify major or radical innovations in technology since then, except for biotechnology.

The surges in invention and technological development which marked the First, Second and Third Industrial Revolutions coincided with periods of intense intellectual, cultural and political activity. The period 1780–1840 included the American Revolution, the French Revolution, the Napoleonic wars, wars of liberation in Latin America and the proclamation of the Monroe Doctrine; a shift in economics from mercantilism to free trade; and the transition from Classicism to Romanticism in music, literature and art (e.g. Mozart, Haydn, Beethoven, Schubert, Chopin, Berlioz, Goethe, Schiller, Heine, Blake, Wordsworth, Turner, Constable, Delacroix, Goya).

The period 1860–1910 included the unification of Italy and Germany, the Meiji Restoration in Japan, the American Civil War, the Franco-Prussian War, 'new imperialism' and intense colonial rivalry; universal male suffrage; Darwin's *On the Origin of Species*; Marx's *Capital*; the laws of thermodynamics, genetics, geology, aerodynamics and molecular structure; Mendeleev's 'periodic' table of the elements; discovery of x-rays and radioactivity, the germ theory of disease (Pasteur); the high point of Romanticism in music (Wagner, Verdi, Brahms, Tchaikovsky, Mahler, Strauss); the writings of Dickens, Flaubert, Hugo, Tolstoy, Dostoevsky, Melville, Ibsen, Kipling; Impressionism, Post-Impressionism and Expressionism in painting (Manet, Monet, Renoir, van Gogh, Cézanne, Seurat, Gauguin, Degas, Munch); *la Belle Epoque* in France (Rodin, Proust, Debussy, Ravel, Stravinsky, Matisse, early Picasso); universal primary education; pioneering psychoanalysis (Freud); Einstein's theory of relativity; and the pioneer film-makers (Lumière brothers and Meliès).

The three industrial revolutions

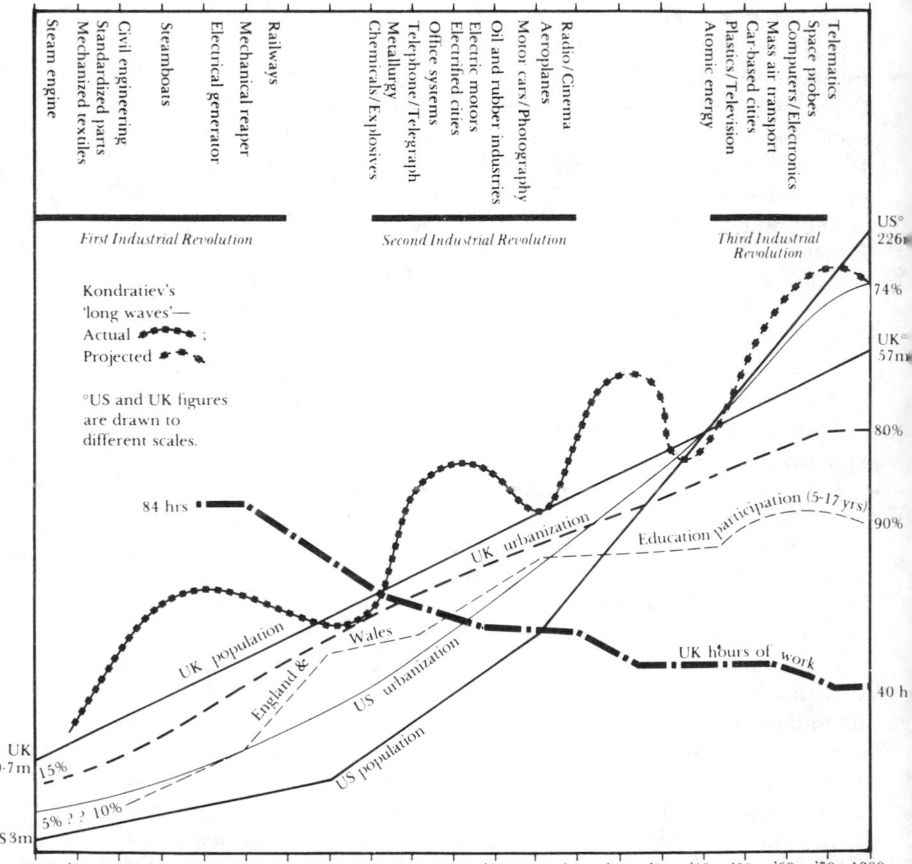

The period 1942–70 included World War II, the first atomic weapons, the setting-up of the United Nations, Soviet domination of Eastern Europe, Indian independence (and partition), the Chinese Revolution and Maoism, self-government in Africa and South-East Asia, establishment of the 'Third World' concept; the development of penicillin, antibiotics and DDT; 'women's liberation', the contraceptive pill, and increased female participation in the paid labour force; Keynesian revolution in economics; rise of multi-national corporations and growth of a global economy; great achievements in painting and sculpture (e.g. Picasso, Moore, Rothko) and architecture (e.g. Le Corbusier, Mies, Aalto, Tange, Johnson); music (e.g. Shostakovich, Britten), and film-making (e.g. Welles, Bergman, Buñuel, Kurosawa).

KONDRATIEV'S 'LONG WAVES'

In 1926 the Russian economist Nikolai D. Kondratiev published an important article, 'The Long Waves in Economic Life', in which he drew attention to significant cyclical variations in economic history.[3] After examining wholesale prices, interest rates, wage levels, foreign trade, and consumption of coal, pig-iron and lead since the onset of the Industrial Revolution, Kondratiev identified two and a half 'long waves of cyclical character' between 1789 and 1920 in Europe and the United States. The first wave, 1789–1849, was on the upswing until 1814 and on the downswing from then until 1849; the second, 1849–96, rose until 1873 and then fell; Kondratiev identified only the upswing of the third wave, 1896–1920. (By projecting future 'long waves' on the basis of a fifty-year cycle – something Kondratiev did not attempt – 1920–45 would be years of decline, 1945–70 years of upswing, followed by a decline 1970–95.)

Kondratiev concluded that 'important discoveries and inventions' were made during the downswing of the long waves but were

> usually applied on a large scale only at the beginning of the next long upswing . . . It is during the period of the rise of the long waves, i.e. during the period of high tension in the expansion of economic forces, that, as a rule, the most disastrous and extensive wars and revolutions take place.[4]

Kondratiev thought that the causes of long waves were 'inherent in the essence of the capitalist economy', but had 'no intention of laying the foundations for an appropriate theory of long waves'.

After the economic downturn of the 1930s Joseph Schumpeter, Arthur Burns and Colin Clark drew attention to Kondratiev's work; and the recession of the 1970s stimulated even more interest in 'long waves'. Simon Kuznets, Nobel Prizewinner in Economic Science for 1971, proposed a modification of Kondratiev's hypothesis, pointing out that in alternating decades the United States had appeared to prosper (1880s, 1900s, 1920s, 1960s) and to regress (1870s, 1890s, 1930s, 1950s, 1970s).

I have tried to reconcile Kondratiev's long waves with the periods of inventiveness in the three Industrial Revolutions, but they do not fit convincingly (e.g. the Electric Revolution covers the whole of Kondratiev's second wave and part of the upswing on the third). They do, however, illustrate the distinction between absolute and relative growth. If the 1970s are regarded as part of a (projected) Kondratiev downswing and represent a decline relative to 1960s growth rates, it must be remembered that in absolute terms production and consumption are at almost their highest recorded levels.

FACTORS IN EMPLOYMENT GROWTH

The dramatic expansion of advanced economies in the past two hundred years is the result of six closely linked factors:

1 *Population growth.* The population of the United Kingdom (England, Wales, Scotland and Ulster) was 9·7 million in 1780 and 57 million in 1980 (nearly a sixfold increase). The US population has grown by a multiplier of seventy-five in the same period (from 3 million to 226 million); and Australia's by a multiplier of seventy-two (from 200 000 – all Aborigines – to 14·4 million). The main reason for growth was the fall in infant mortality and the diseases of children: the life span of mature adults has only increased by a decade since 1800.

2 *Urbanization.* In 1780 about 15 per cent of people in the United Kingdom, 5 per cent in the United States, and 0 per cent in Australia, lived in cities or towns: two hundred years later the figures were 80 per cent, 74 per cent and 86 per cent respectively. In 1780 there were only five cities (Tokyo, and four in China) throughout the world with a million people or more: by 1980 there were 160.

3 *Division of labour and increasing complexity of transactions.* As cities grow, transactions become increasingly complex. Services have to be provided for entire communities, leading to a decline in self-sufficiency. As a result of specialization, individuals devote themselves to a limited range of functions which are then in effect 'exchanged' – generally through cash transactions – for somebody else's specialist products. Urbanization and division of labour are inextricable: they rise and fall together. When a farmer takes an egg from his own hen and cooks it himself, he is the only person in the transaction. City egg-consumers are at the end of a long production chain which involves farmers, carton-makers, truck-drivers, marketing-board administrators, insurers, storemen, date-stampers, invoicers, salespeople and cooks.

4 *Technological change.* Technological change is both a result and cause of population growth, urbanization and division of labour.[5] The growth of cities and the building of factories and shops require new construction techniques and reliable transport systems. Changes in medicine and public health (especially clean water supply) have led to dramatic falls in mortality rates. The form of technology chosen, e.g. private cars instead of trains and buses, changes patterns of urban living. High-volume production allows for lower unit costs for goods and increased consumption.

5 *Changes in working patterns.* In pre-industrial societies, work was governed by the sun and the seasons, not by the clock; there is no

precise record of how much labour was expended each day, week or year. After the Industrial Revolution urban workers toiled for up to eighty-four hours each week, certainly far longer than their parents or grandparents. The Cotton Mills Act (1819) set a maximum working week of seventy-two hours for children; for their parents the alternative to long working hours was unemployment and destitution. With increasing industrialization the hours, days, weeks and years of employment were gradually reduced and available work was shared: the 40-hour week became standard in Europe and North America after World War II.

6 *Rising levels of education and literacy.* Educational participation rates and literacy rates have been the most reliable indicators of rising living standards in the past two hundred years, and major factors in encouraging high-volume production and diversity in employment.[6]

HOW TECHNOLOGY CHANGES WORK: THE HISTORICAL EXPERIENCE

Historically, the impact of technology has been to increase productivity in specific areas and, in the medium or long term, 'release' workers who can be used in the creation of other forms of work. For example, the new technologies applied to British farming in the seventeenth and eighteenth centuries (and ever since) permitted agricultural employment to fall from 75 per cent of the total labour force in 1688, to 50 per cent before 1780, 25 per cent by 1840 and 3 per cent by 1980. Between 1780 and 1980, the agricultural labour force was reduced by eleven-twelfths (in absolute figures, the reduction was by one-half), while productivity per capita increased by a factor of 68 to one. This created opportunities for work expansion in *other* areas – and the early nineteenth century was marked by a rapid increase in employment in both industry and services. There was a movement between sectors like this:

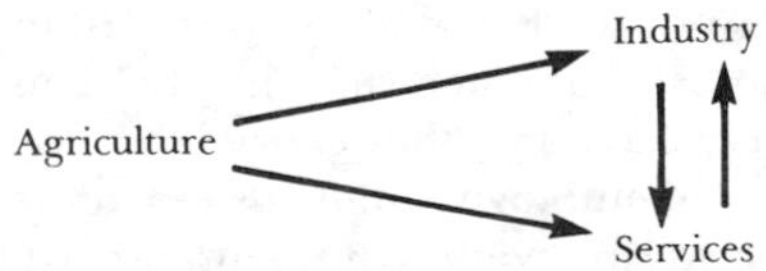

During the First Industrial Revolution there was a set of interacting causes and effects, of which the most significant elements (and the approximate order in which they occurred) were:

1 Elimination of subsistence agriculture and decline in craftwork.
2 Growth of industrial 'process' work (including machine-minding).
3 Increase in child, female and part-time labour as a proportion of the total labour force.

4 Reduction in wages relative to value of total output.
5 Rapid growth in employment outside the technologically affected areas (because high unemployment made people available for work at lower rates).

During the Second Industrial Revolution, technological innovation led to major changes in employment:

6 Expansion of female employment based on labour-intensive information transfer (education, office work), and male employment in making and using motor vehicles.

In the British cotton industry, the application of the spinning jenny (1770), the water frame (1773), the spinning mule (1779), the power loom (1785), steam power (1785) and the self-acting mule (1825) led to a hundredfold increase in output between 1760 and 1827. Increases in employment, however, were far more modest: machinery transformed many workers from craftsmen into machine-minders, the numbers of skilled adult males fell sharply relative to output (although probably increasing absolutely), and the domestic system of 'putting out' work was replaced by employment in factories. The combination of Whitney's cotton gin and enormously increased demand for cotton led to a rapid growth in Negro slavery in the USA – from 1·5 million in 1820, to 4 million forty years later. As late as the 1891 census, the largest single occupational groups in Great Britain were 'agriculture, horticulture and forestry' and 'domestic offices and personal services'.[7] Workers in textiles and clothing were treated as separate categories: if taken together, they declined from 23·5 per cent of the labour force in 1851 to 15 per cent in 1911. (Domestic servants were stable at 14 per cent in both censuses.) As technology was applied to domestic service with vacuum cleaners, washing machines and driers, employment in that sector declined to create opportunities for expansion elsewhere.

Some new forms of work (e.g. railways in Britain in the nineteenth century, and motor manufacturing in the United States in the twentieth century) have in the short and medium term been enormous creators of employment. British locomotives, carriages and rails were exported to every continent, as were US Fords and Chevrolets: as 'headquarters economies' for the entire world, it was possible for Britain and the United States to have high productivity *and* high employment in these industries. If they had been only providing for a national market, employment in manufacturing would have peaked and fallen decades earlier than it did. Technology *always* changes pre-existing employment. The pattern of change is not:

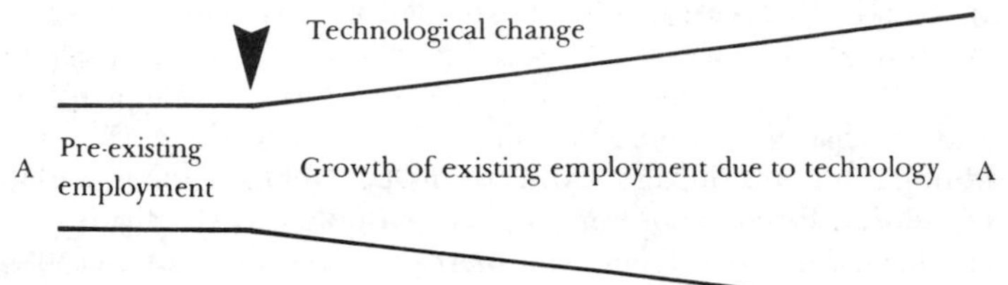

Historically, the major long-term effect of technological change is a 'transfer' – the generation of work which is *complementary* to the areas technologically affected.

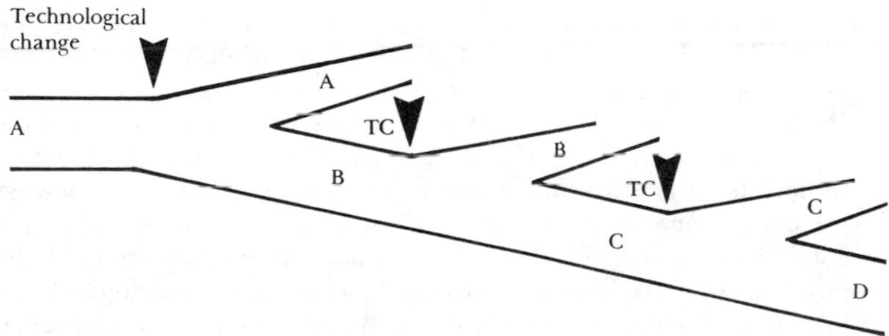

If past analogies are the best guides to the future, we may assume that as technology revolutionizes information-based employment then that type of employment will contract and not expand, leading either to the creation of new types of non-technologically based work or to the growth of a 'reserve army of the unemployed'. This phenomenon is strikingly confirmed by US labour-force figures, which indicate that between 1970 and 1980 – a time of rapid computerization and expansion of information sources – paid employment in 'information' (median definition, see also Chapter 3) increased from 46·4 per cent to 46·6 per cent, the smallest recorded growth rate for any decade. Similarly, Australian estimates for 'information' employment indicate a slight downturn between 1976 (30·9 per cent) and 1979 (30·8 per cent).

US Department of Labor statistics indicate that in the 1970s the greatest rate of employment growth was in relatively servile or poorly paid service employment: the eating and drinking industry (including fast-food chains), care of the aged and sick, building maintenance, cleaning and delivery services.

The social impact of the First and Second Industrial Revolutions

Vast increases in pauperism after 1780 led to the breakdown of traditional systems of parish- and county-based 'poor relief', and rapid urbanization forced adoption of national policies. The British Parliament passed the 1834 Poor Law Act, in the spirit of Benthamite utilitarianism, with the deliberate intention of forcing people to seek work – however degraded and wretchedly paid – rather than seeking 'charity' in 'workhouses', a situation exposed by Dickens in *Hard Times* (1854). E. P. Thompson quotes an Assistant Commissioner of the Poor Law: 'Our intention is to make the workhouses as like prisons as possible . . . and to establish therein a discipline so severe and repulsive as to make them a terror to the poor and prevent them fom entering.' Families were segregated, their possessions taken away, and inmates were subject to a strict routine of silence and religious exercises designed to eliminate individualism and the seeds of rebellion. By 1843, the workhouses held 197 179 inmates. 'The most eloquent testimony of the depths of poverty is in the fact that they were tenanted at all.'[8]

Child labour was intensified and institutionalized between 1780 and 1840, and then diminished slowly. It was abolished in Britain by the Children and Young Persons' Act (1933) and in the United States by the Fair Labor Standards Act (1938). Under the domestic system of industry children had usually shared work with their parents, but they had a variety of small tasks to break the monotony, tempered by both play outdoors with other children and family affection. Even though male wage rates fell dramatically after 1810, employers still preferred to use part-time rather than full-time workers, and women and children rather than men. (In 1838, only 23 per cent of textile workers in Britain were adult males.) When children worked in factories and mines they were governed by overseers and by the clock: they never saw the sunlight at work, and the tasks lacked variety. Discipline was severe – safety regulations were not. In 1842, Parliament prohibited the employment of women or children under ten in the mines, while in 1847 children were limited to ten hours' work per day in textile factories. The last elements of the 1834 Poor Law were not abolished until 1929 – just before the greatest employment slump for a century.

In February 1811, a wave of machine-breaking began in the hosiery and lace industry in the adjoining counties of Nottingham, Leicester and Derby, and continued intermittently until April 1812 by which time 1000 knitting frames had been smashed. The machine-breakers described themselves as 'Luddites' in proclamations published in the name of the legendary 'General Ned Ludd'. Luddite activity also spread to the woollen mills of Leeds, Bradford, and Halifax in Yorkshire, then to the croppers (cloth finishers) and cotton-weavers in Lancashire and Cheshire who opposed the re-introduction of the gig-mill (use of which had been restricted by an Act of Edward VI in the sixteenth century). Frame-breaking was made a capital offence in 1812. Many Luddites were hanged for various acts of violence, including murder, and more were transported and jailed. There were three comparatively minor outbreaks of machine-breaking in 1812–13, 1814 and 1816, but by 1817 Luddism had disappeared (although many Luddites became Chartists).

The Luddites were skilled workers who looked backwards to medieval craft traditions and forwards to modern trade unionism. They put heavy emphasis on the importance of maintaining high-quality products, and their targets were very carefully chosen. Most craft workers still worked under the domestic system, which meant that striking was not appropriate, and there were no other means to express their frustration. They were opposed not to machines as such, but rather to the 'de-skilling' involved in destroying craft industries: they concluded that the new technology would destroy their livelihood and opportunities for work. It did. Forty years later, in 1845, the Midlands hosiery and lace industries were still depressed despite the rosy optimism of the Industrial Revolution's contemporary advocates.[9] E. J. Hobsbawm has pointed to

> the waves of desperation which broke time and again over [Britain]: in 1811–13, in 1815–17, in 1819, in 1826, in 1829–35, in 1838–42, in 1843–44, in 1846–48. In the agricultural areas they were blind, spontaneous, and, in so far as their objectives were at all defined, almost entirely economic.[10]

The term 'Luddite' is invariably used to describe mindless obscurantism and machine-breaking. It would be better applied to intelligent industrial activists, who want to bargain about the *terms* on which new technology should be introduced – i.e. What forms? How much? When? With what social effects? Under whose control? Are there alternatives? Who receives the benefits? Lewis Mumford, after considering the role of the Luddites, went on:

> What shall we say of the counter-Luddites, the systematic craft-wreckers ... the ruthless enterprisers who, during the last two centuries, have in effect confiscated the tools, destroyed the independent work-shops, and wiped out the independent tradition of handicraft culture? What they have done is to debase a versatile and still viable polytechnics to a monotechnics and at the same time they have sacrificed human autonomy and variety to a system of centralised control that becomes increasingly automatic and compulsive.[11]

The First Industrial Revolution produced massive and long-lasting structural employment and instability. Intelligent state action could have greatly increased both the humanity and the business efficiency of the transformation.

In the turbulent period 1815–50, there were four distinct responses to the impact of industrialization and urbanization in Europe – conservative (attempts to restore the *status quo ante* after the French Revolution and Napoleon), technological (Saint-Simon's plans for a rational society based on science), nationalist (Mazzini's programme for an international federation of national states – taken up by 'Young Poland', 'Young Italy' and 'Young Ireland') and communist (Marx's advocacy of an international revolution by the industrial proletariat).

The Second Industrial Revolution was marked by the phenomenon of 'new imperialism', which began rather suddenly after 1873–74 when Britain, the newly united Italy and Germany, France (anxious to recoup prestige after her losses to Germany), and Belgium began systematically colouring in blank areas on the maps of Africa, China, South-East Asia and the Middle East. The United States extended its sphere of influence throughout the Pacific and Latin America. There were a dozen contradictory strands in new imperialism: economic exploitation and economic uplift, the lure of cheap labour and anti-slavery zeal, the need for military bases and the desire to extend Christianity, fear, greed, social Darwinism ('the survival of the fittest'), and the extension of European culture. The voracious demand for raw materials, cotton, rubber, tin and oil was paramount. The rivalries generated by colonial competition were major factors in the rise of nationalism, massive expenditure on armaments, and the increase in international tension which contributed to the outbreak of World War I in 1914, ending a century of comparative peace in Europe since Waterloo.

LAISSEZ-FAIRE VERSUS INTERVENTIONISM

The prevailing economic philosophy of the First Industrial Revolution was the concept of *laissez-faire* – in effect, 'leave things alone', 'let things happen', 'leave it to the market' – first proposed by the French physiocrats and then propounded elegantly and persuasively by Adam Smith in *An Inquiry into the Nature and Causes of the Wealth of Nations* (1776).[12] This three-volume work set out the economic philosophy underlying the Industrial Revolution and modern capitalism. Smith argued that individual self-interest created an 'invisible hand' which would optimize wealth and indirectly create benefits for all:

Every individual . . . endeavours as much as he can . . . to direct . . . industry that its produce may be of the greatest value; every individual necessarily labours to render the annual revenue of the society as great as he can. He generally, indeed, neither intends to promote the public interest, nor knows how much he is promoting it. . . By directing that industry in such a manner as its produce may be of the greatest value, he intends only his own gain, and he is in this, as in many other cases, led by an invisible hand to promote an end which was no part of his intention. Nor is it always the worse for the society that it was no part of it. By pursuing his own interest he frequently promotes that of the society more effectually than when he really intends to promote it.

He also developed what was later called 'equilibrium theory' (based on analogies from physics), the concept that supply and demand automatically adjust to find their own level through the operation of price competition. (The price of labour – i.e. wages – is subject to the same effects.) Accordingly the 'market' is seen as the best mechanism for determining both what goods are produced and their appropriate price (given, as Smith said, that there is perfect competition and perfect information): the state should not interfere, because this would distort the market and ultimately cause disbenefits. Smith notwithstanding, the First Industrial Revolution in Britain – although apparently based on *laissez-faire* and self-interest – was protected by massive, though probably unnecessary, tariffs and by heavy and very helpful public expenditures throughout the Napoleonic wars on roads, docks, ships, armaments, uniforms and equipment. The Australian historian Dr F. B. Smith has conjectured that new public investment may have exceeded new private investment around 1806: free trade and unaided private enterprise did not produce the steam engine and all its wealth.

Smith's ideas were developed by the French industrialist and economist Jean-Baptiste Say (1767–1832). 'Say's law' (implicit in his works but not explictly stated) declared that increased supply of goods creates its own demand, increased demand (in its turn) leads to increased supply, and that there are, in effect, no limits to employment growth. Say's law has been revived in the United States as an essential element of 'supply side economics', as advocated by George Gilder and David Stockman. This asserts that if the poor were taken off welfare, with its emphasis on 'demand side economics', and borrowed enough capital to set themselves up shining shoes or selling flowers in the street, their 'supply side' efforts would generate considerable demand thus stimulating a shoeshine-led economic recovery. In Britain, David Ricardo (1772–1823) and the 'Manchester school' (as Disraeli derisively called John Bright, Richard Cobden and their followers) argued for free trade and against any forms of government restraint or encouragement in economic matters. After 1846, when the Corn Laws were abolished, Britain pursued free trade for eighty-six years – a period in which she continuously lost competitive advantage to her industrial rivals (notably Germany and the United States), who had come to understand during the Second Industrial Revolution the services which states could provide to assist capitalist growth. The resort to free trade and small government as advocated by Gladstone was adopted only by Britain, and proved generally disastrous in the long term. Both parties acquiesced in free trade until Joseph Chamberlain began to campaign vehemently for tariffs (1903–06): in 1932 his son Neville introduced 'Imperial Preference' as a form of protection against the Great Depression.

In the United States, Washington's Secretary of the Treasury, Alexander Hamilton, while accepting much of Smith's approach in *Wealth of Nations*, rejected *laissez-faire* and in his important *Report on Manufactures* (1791) urged government support and tariff protection for infant industries – a policy which was adopted during the Second Industrial Revolution.

The most passionate opponent of *laissez-faire* was Karl Marx. Born in Germany in 1818, Marx was exiled from his homeland after 1843 and again from 1849, living in London from 1850 until his death in 1883. His formative years were influenced by the First Industrial Revolution – especially the division of labour, the end of the domestic system, the concentration of labour in central workplaces, and the introduction of industrial discipline with its accompanying alienation, boredom, and fatigue. Marx wrote *The Communist Manifesto* in 1848, and his masterpiece, *Capital*, was published in three volumes (1867, 1885 and 1894). He attempted to do for history what Darwin had done for biology. Marx argued in *Capital* (Volume I) that demand for labour falls

> progressively with the growth of the total capital . . . It falls relatively to the magnitude of the total capital, and at an accelerated rate, as this magnitude increases . . . But if a surplus population of workers is a necessary product of accumulation or of the development of wealth on a capitalist basis, this surplus population also becomes, conversely, the lever of capitalist accumulation . . . It forms a disposable reserve army . . . a mass of human material always ready for exploitation by capital . . . The general movements of wages are exclusively regulated by the expansion and contraction of the industrial reserve army.[13]

The rise of what was historically a new class, the industrial proletariat, led to the slow development of trade unions. The year of revolutions – 1848 – was partly caused by economic turndown and high unemployment in the 1840s, and failed because the industrial labour force was still tiny and poorly organized. The growth of trade unions and socialist, democratic or populist parties during the Second Industrial Revolution represented attempts to regulate the economy and ensure more equitable sharing of resources, and to urge the state to protect the interests of the weak against the strong. British unions were given rights of legal incorporation in 1871, and French unions in 1884. Unionism remained very weak in Europe and North America until after 1900.

As already noted, during the Second Industrial Revolution the governments of France, Germany and Russia used interventionist techniques to stimulate and regulate the economy, in complete contrast to Britain. France had begun to industrialize in 1790, but after the Napoleonic wars ended in 1815 the political system was dominated by conservative rural interests who discouraged urban growth and recalled the revolutionary role of Paris, Lyon and Marseille with horror: industrial development remained comparatively immature until after 1860. Napoleon III instituted price controls, legalized trade unions, encouraged public works, constructed state-owned railways, subsidized workers' housing, and promoted inventions.

In Germany, Bismarck – following the economic theories of Friedrich List – encouraged the formation of cartels or state-sponsored monopolies to control heavy industry, while the introduction of high tariff protection at home guaranteed enough profit to allow exporters to undercut foreign competition. Germany rapidly passed Britain in some manufacturing areas. Between 1883 and 1889, Bismarck laid the foundation of a welfare state by providing for sickness and accident insurance (subsidized by employers) and old-age and invalid pensions (subsidized by the state). He also recognized the link between education and productivity.

In Russia, Count Witte (Finance Minister 1893–1903) forced the pace for industrial development, encouraged large-scale factories, built 100 000 kilometres of railway track, and tried to introduce business principles into government. Russia's late industrialization meant that she acquired technology in an advanced form and went directly into high-volume production in large factories.

Even in Britain, the state intervened in some economically related areas: safety regulations, maximum hours of work for women and children (none of these was rigorously enforced) and, in 1870, universal primary education. Most German states had introduced compulsory schooling long before industrialization (e.g. Austria in 1783), and Britain's Board of Trade sent a mission to Germany in 1868 which reported that competitive public education was an essential precondition to competitive private industry. Schooling in England had hitherto been left to churches and voluntary organizations, but the Education Act of 1870 made primary education a national responsibility and attendance until the age of ten became compulsory. The law did not come into effect until 1876, and its enforcement was spasmodic and half-hearted; the leaving age was raised to eleven in 1893, and twelve in 1899 (compulsory secondary education to the age of fifteen was not established until the Butler Act of 1944). The 1870 Act was passed for utilitarian reasons – instilling order and discipline so that the future labour force could learn to obey orders without hesitation, while literacy and numeracy had practical benefits in improving quality of work. In the United States, education had long been a state matter. Between 1830 and 1870, locally elected school boards were set up in most areas to provide secular education for all white children and most black ones. France introduced compulsory primary school attendance in 1882.

The *laissez-faire* versus interventionism debate is as significant, and unresolved, today as it was a hundred years ago. Milton Friedman and his monetarist supporters – including such conservative politicians as Margaret Thatcher, Ronald Reagan, Malcolm Fraser and Menachem Begin, and the Australian economists Richard Blandy and Wolfgang Kasper – argue that capitalism is essentially self-regulatory, and urge the highest degree of government disengagement from the economy as is politically possible.[14] (Dismantling of the welfare state is not a realistic option.) The followers of J. M. Keynes (currently unfashionable) argue that free-ranging market forces are not enough to ensure the efficient and equitable working of the economy, and that governments should intervene where appropriate by using macro-economic techniques such as varying exchange rates, tariffs or taxes, or by raising (or lowering) the volume of money in circulation. These techniques have been used by Labour or Social Democratic governments in Britain, Australia, Germany and Scandinavia, and by liberal Democrats in the United States.

It can be argued that, in the long term, economic reliance on an all-powerful state has caused greater trauma than has a tradition of scepticism and detachment about government intervention. Hitler was uniquely successful in providing full employment in Germany, and Stalin's Russia was largely insulated from the Great Depression – but unresisted statism created horrors of a different nature. On the other hand, the free market has never provided full employment except in wartime. Both *laissez-faire* and interventionism have had massive failures – neither has combined economic efficiency, social efficiency and the growth of human dignity.

Unemployment and the Industrial Revolutions

Since the First Industrial Revolution began in Britain in 1780,

US unemployment, 1900–63

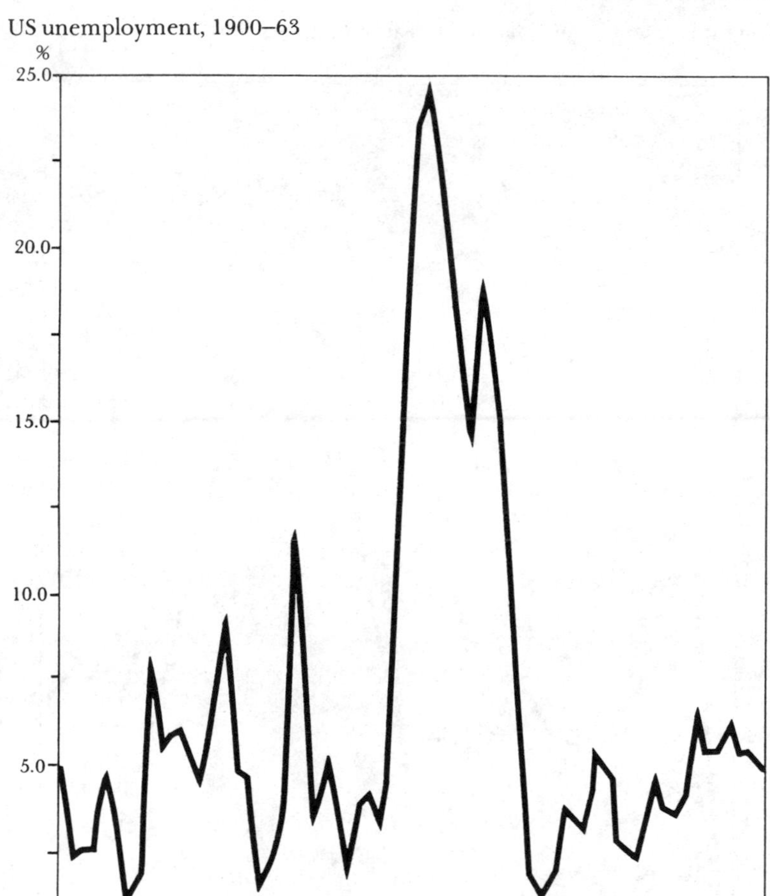

Source: Paul A. Baran and Paul M. Sweezy, *Monopoly Capital* (Penguin, London, 1977), p. 229.

unstable employment has been the norm. Since then there have been thirty years of full employment (1945–74), forty years of war, and 130 years of unstable employment with sharp alternations between high and low levels.[15] Nevertheless, optimists often argue as if employment statistics only began in 1945, that full employment has been the norm, and that increasing joblessness since 1974 has been a temporary aberration soon to be ended by a revival of consumer demand. Given the evidence of the past two hundred years, this is an extremely dubious proposition.

Unemployment in Australia and Great Britain, 1920–39

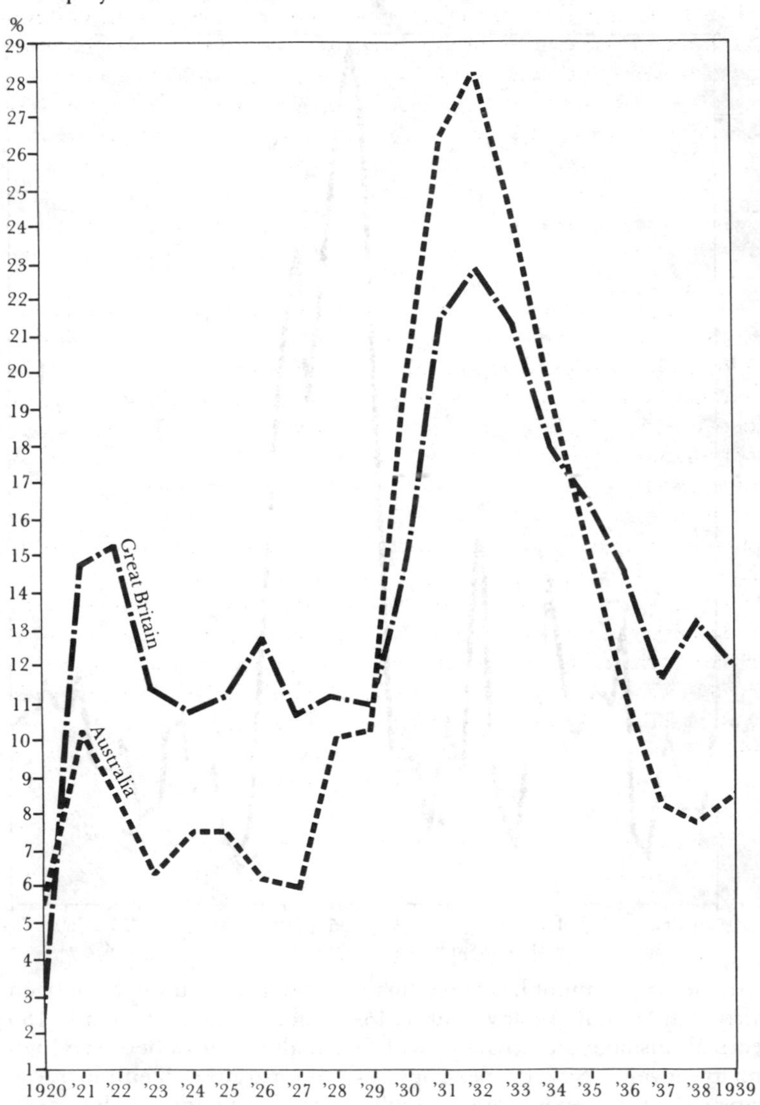

Source: (for the British figures) B. R. Mitchell and Phyllis Deane, *Abstract of British Historical Statistics*, pp. 65, 67.

The term 'unemployment', used to describe a social situation rather than a personal condition (*idleness*), first appeared in trade-union, radical or Owenite writings in the 1820s and 1830s.[16] Unemployment figures were not recorded in Britain until the 1860s. In the United States self-employment (and near-destitution) was the dominant work mode until the Civil War, so 'unemployment' was a minor problem. Statistics for unemployment are notoriously unreliable and are rarely standardized for international comparison. British and Australian figures were recorded by trade unions until the 1920s, so that non-unionists were not counted; under-employment was not measured, either.

Average unemployment in the United States has been estimated as 10 per cent in the 1870s and 1890s, and 4 per cent in the 1880s and 1900s. The figures for 1900–63 are given in the graph on p. 27. The extent of unemployment in Australia and Great Britain in the period between World Wars I and II – when Great Britain still had a commanding position in international trade – is shown in the graph on p. 28. (Up to 1923 the base figures for Great Britain are those from trade unions recording unemployment against their membership; thereafter they relate to workers covered by National Insurance. Neither set of figures is comprehensive, and the true global rate of unemployment at any time was probably somewhat higher than indicated.) The most comprehensive international comparison of unemployment rates was for the same period and appears in *The Measurement and Behaviour of Employment*, published by the Universities-National Bureau Committee for Economic Research (Princeton, 1957, pp. 455–6) and is shown on p. 30. Contemporary unemployment, and its relationship with inflation and productivity, are considered in Chapter 6.

Keynes rejected the equilibrium theory in relation to employment. He proposed strict government control of credit to restore confidence to the economy, stimulate investment, increase social services, build public works, and encourage consumption to increase employment. (The Polish economist Michal Kalecki [1889–1970] anticipated Keynes' theories on demand management and employment, but his priority was only acknowledged after Keynes' death.) In the 1930s, Adolf Hitler's 'New Order' in Germany, Franklin D. Roosevelt's 'New Deal' in the United States, and Per Albin Hanssen's Social Democratic government in Sweden adopted these interventionist policies with varying success. World War II created domestic 'full employment'. Great Britain adopted Sir William Beveridge's White Paper entitled 'Full Employment in a Free Society' (1944); Keynes' ideas were the

Unemployment rates, ten countries, 1920–40 (per cent)

Year	Australia	Belgium	Canada	Denmark	Germany	Netherlands	Norway	Sweden	United Kingdom	United States
1920	5·5	–	4·6	6·1	3·8	5·8	2·3	5·4	3·2	4·0
1921	10·4	9·7	8·9	19·7	2·8	9·0	17·7	26·6	17·0	11·9
1922	8·5	3·1	7·1	19·3	1·5	11·0	17·1	22·9	14·3	7·6
1923	6·2	1·0	4·9	12·7	10·2	11·2	10·7	12·5	11·7	3·2
1924	7·8	1·0	7·1	10·7	13·1	8·8	8·5	10·1	10·3	5·5
1925	7·8	1·5	7·0	14·7	6·8	8·1	13·2	11·0	11·3	4·0
1926	6·3	1·4	4·7	20·7	18·0	7·3	24·3	12·2	12·5	1·9
1927	6·2	1·8	2·9	22·5	8·8	7·5	25·4	12·0	9·7	4·1
1928	10·0	0·9	2·6	18·5	8·6	5·6	19·2	10·6	10·8	4·4
1929	10·2	1·3	4·2	15·5	13·3	5·9	15·4	10·2	10·4	3·2
1930	18·4	3·6	12·9	13·7	22·7	7·8	16·6	11·9	16·1	8·9
1931	26·5	10·9	17·4	17·9	34·3	14·8	22·3	16·8	21·3	15·9
1932	28·1	19·0	26·0	31·7	43·8	25·3	30·8	22·4	22·1	23·6
1933	24·2	16·9	26·6	28·8	36·2	26·9	33·4	23·3	19·9	24·9
1934	19·6	18·9	20·6	22·2	20·5	28·0	30·7	18·0	16·7	21·7
1935	15·6	17·8	19·1	19·7	16·2	31·7	25·3	15·0	15·5	20·1
1936	11·3	13·5	16·7	19·3	12·0	32·7	18·8	12·7	13·1	17·0
1937	8·4	11·5	12·5	21·9	6·9	26·9	20·0	10·8	10·8	14·3
1938	7·8	14·0	15·1	21·5	3·2	25·0	22·0	10·9	12·9	19·0
1939	8·8	15·9	14·1	18·4	0·9	19·9	18·3	9·2	10·5	17·2
1940	7·1	–	9·3	23·9	–	19·8	23·1	11·8	5·0	14·6

basis of policies agreed at the Bretton Woods Conference in 1944, where the International Monetary Fund was set up.

The 'golden age' of full employment in peacetime ran from 1945 to 1974 in Britain, the United States, Australia, Canada, New Zealand and Scandinavia. (It is fair to note that Japan maintained full employment without adopting Keynesian policies, partly due to the retirement age of fifty-five and a lower rate of female participation in the labour force.) The welfare state was promoted by Clement Attlee's Labour government in Britain, Harry Truman's Fair Deal in the United States, Ben Chifley's Labor government in Australia and Tage Erlander's Social Democrats in Sweden. There was an explosive increase in secondary and tertiary education, the development of a motorized society with huge investments in car manufacturing and road construction, rapid rates of population increase and urban growth, tariff protection, subsidies for industry (where appropriate), and high levels of defence expenditure during the Cold, Korean and Vietnam wars, intensified by the nuclear arms race.

Britain was a pioneer of the 'mixed economy' in which capitalism and nationalized industries co-existed – and where governments had a decisive influence on profitability in the private sector by controlling interest, taxation or wage levels. Despite political rhetoric by advocates for socialism or unreconstructed capitalism, pragmatists in parliament, business and bureaucracy held the middle ground and (until Mrs Thatcher's victory in 1979) accepted that there would be only minimal moves away from public ownership towards free enterprise, or vice versa. (This consensus approach was sometimes called 'Butskellism' after R. A. Butler [Tory] and Hugh Gaitskell [Labour], Chancellors of the Exchequer whose economic views converged.)

If, as economic optimists urge, the best guide for the future is the past, then it should be pointed out that in the past two hundred years of economic history unstable employment has been the norm, and market forces alone have never succeeded in establishing full employment without government assistance.

The case for continuity and optimism

Economic optimists rely on Adam Smith and equilibrium theory, insisting that market forces regulate employment. Thus, in times of labour scarcity wages (the price) will go up, and in times of labour abundance wages will fall. Providing that market distortions – such as maintaining high wages in times of unemployment – are avoided, there will in the long term be adequate employment for all who want it. There may be temporary or 'frictional' unemployment (e.g. in periods of transition from one job or area to another), but not long-term structural or technological unemployment.

Economic history is neither a fashionable nor a rigorous discipline. It is commonplace for optimists to assert that in economics there is nothing new under the sun – no new problems, only old recurring ones which are triumphantly overcome. Thomas Carlyle called economics 'the dismal science' – but this only revealed his ignorance of economists.[17] They are as a group the most irrepressible Micawbers to be found anywhere: assertions that the present era is not one of revolutionary change or discontinuity with past economic history are made with an astonishing assurance, in inverse proportion to the quantity (or quality) of historical analysis presented. Typically, assumptions are made and repeated at length, but no historical evidence is presented. The credo is a matter of justification by faith alone; no documentation is possible or apparently considered necessary.

An expensive example of this is found in the four-volume report of the Committee of Inquiry into Technological Change in Australia (CITCA), also known as the Myers Report.[18] The committee, consisting of a metallurgist, engineer and union secretary, and serviced by the Department of Productivity, was appointed in December 1978 to report on 'the process of technological change in Australian industry' – broadly defined – 'to maximize economic, social and other benefits and minimize any possible adverse consequences'. The central argument of the report – that employment prospects under technological change should be viewed optimistically, and that the present era represents continuity with the past – is set out as follows in Volume I:

> [Para. 4·57] The Committee ... does believe that the available historical evidence shows that technological change has in the long term created wealth and employment and that future technological changes will continue to have this effect.[19]

This sentence comprises the committee's complete analysis of a complex and controversial issue, although paragraphs 4·20, 4·21 and 4·22 provide brief lists of developments in the Neolithic Revolution, the Industrial Revolution and the adoption of the deep plough in Europe about 1000 AD – less than one and a half pages in a total of 1556. The assertion begs four questions:

1 Why is the evidence of the past regarded as the best guide for the future?

2 What evidence is there to suggest that the present era represents continuity rather than discontinuity with past economic history? If there is any, where is it examined in the report?

3 How long is the 'long term' – ten years, twenty years, fifty years? The period itself is critical to appropriate social planning.

4 Does technological change produce increased or reduced employment in the specifically affected areas?

The report urged rapid adoption of technology, recommending consultation with unions and a 'social safety net' to provide compensation for workers displaced or retrenched by technology. The Fraser government adopted the bulk of the report but rejected the 'social safety net'. In announcing the government's response to the report, Sir Phillip Lynch, Minister for Industry and Commerce, argued the case for continuity and optimism with two curious examples:

> Innovative technologies also have another very important effect. They frequently create their own demand. The modern jet airliner is a case in point. The rapid spread of national and international air travel could barely have been imagined 20 or 30 years ago. Consumer durables such as television receivers are also examples of significant new industries which have developed as a result of technical innovation.[20]

Aviation is not a rapidly growing employer. The greatest rate of employment growth in Australian aviation took place between 1947 and 1954 (a 95 per cent increase on a low base) and between 1961 and 1966 (a 70 per cent increase). Between 1966 and 1979, the air transport industry grew from 0·50 per cent of the labour force to 0·56 per cent. Further, the idea of international air travel was not just 'barely imagined 20 or 30 years ago' – it has been a staple of imaginative writers since the eighteenth century. Similarly, the Australian electronics industry (including television) has been a disaster area as an employer.[21] In 1970, 35 692 people were employed, but by 1980 the figure had fallen to 20 300. Aviation and television are perceived as the epitomes of modern technology and lifestyle. So they are – but although prodigious producers, they are not great employers.

TOWARDS A POST-INDUSTRIAL ERA

The Third Industrial Revolution was largely a product of World War II, stimulated and prolonged by the Cold War. Government-controlled research teams produced technology of unparalleled power, range, speed and capacity – dominated by atomic weaponry, supersonic aircraft, intercontinental missiles, global communications systems, the space race, computers and micro-electronics. World leadership passed to the two nuclear super-powers, the United States and USSR, until the revival of Europe and the rise of Japan and China. The colonial powers declined, new nations were created in Africa and South-East Asia, and a global economy emerged which led inexorably

Comparison of information society with agricultural and industrial society

	Agricultural society	Industrial society	Information society
Production power structure			
Production power form	Land production power (farmland)	Production power of motive power (steam engine)	Information production power (computer)
	Material productivity	Material productivity	Knowledge productivity
Character of production power	Effective reproduction of natural phenomenon	Effective change of natural phenomenon and amplification	Systemization of various natural and social function
	Increase of plant reproduction	Substitution and amplification for physical labour	Substitution of brain labour
Product form	Increase of agricultural product and handiwork	Industrial goods, transportation and energy	Information, function and system
	Agriculture and handicraft	Manufacturing and service industry	Information industry, knowledge industry and systems industry
Social structure			
Production and human relations	Tying humans to land	Restricting man to production place	Restricting man to social system
	Compulsory labour	Hired labour	Contract labour
Special character of social form	Closed village society	Concentrated urbanized society	Dispersed network society
	Permanent and traditional society	Dynamic and free competitive society	Creative and optimum society
	Paternalistic status society	Social welfare type controlled society	Social development type multifunctional society
Value outlook			
Value standard	Natural law	Materialistic satisfaction	Knowledge creation
	Maintenance of life	Satisfaction of sensual and emotional desires	Pursuit of multiple social desires
Thought standard	God-centred thought (religion)	Human-centred thought (natural science)	Mankind-centred thought (extreme science)
	Ecclesiastical principle	Free democracy	Functional democracy
Ethical standard	Law of God	Basic human rights; ownership rights	Sense of mission and self-control

to the rise of multi-national corporations. While communism was a dynamic force throughout Asia and Eastern Europe, it lost its ideological power after 1970 and became increasingly rigid, conservative and bureaucratic (except in China).

The Third Industrial Revolution permitted the achievement of huge ends by tiny means. Its base was not the production of goods and tangible services, or the elimination of muscle power by machines: it was the collection and dissemination of *information*, which became the central factor in the organization of society and national economies, and promoted the growth of a global economy.

Dr Yoneji Masuda, Director of the Japan Computer Usage Development Institute (JACUDI) and later co-founder of the Institute for Information Society, proposed a succinct comparison of the major differences between pre-industrial, industrial and post-industrial societies.[22] It is reproduced on p. 34. By 1980 the transition from employment dominated by manufacturing and directly related industrial services to employment dominated by post-industrial service occupations has been completed in the United States, the United Kingdom, Canada, Australia and New Zealand, Japan, Sweden, Norway, Denmark and Finland, France, Austria, Italy, Switzerland, the German Federal Republic, the German Democratic Republic, the Netherlands, Belgium and Luxemburg.

Every era raises new problems – but generally rates of change have been slow enough for adaptation to take place over several decades. With the development of a global economy and the rapid diffusion of low-cost micro-electronic technology, the sheer rate of change poses unprecedented problems. The recession which occurred outside the Soviet bloc after 1973–74 was marked by a combination of high inflation, high unemployment and anxiety about reduced rates of economic growth. These elements, partly side-effects of the transition to post-industrialism, coincided with dramatic increases in oil prices set by the Organization of Petroleum Exporting Countries (OPEC). Governments over-reacted to the energy crisis, and failed to recognize the significance of the structural transitions in economies and societies. This delayed recovery.

The 1980s will be marked by a rapid fall in large-scale market-based service employment which is routine or repetitive in nature and can be replaced by computerized operations at lower cost with greater speed and accuracy – office work, banking, insurance, storage, retailing, mail delivery, fuel supply. This will lead to a further transition to a post-service society which is increasingly based on education, home-based employment and leisure industries. But is our era, as conventional wisdom has it, just one more evolutionary stage in the

Industrial Revolution, or is it something fundamentally new – an age of discontinuity?

The case for discontinuity

The following elements relating to the adoption of new technology have no precedent in economic history. In combination they represent a compelling case for discontinuity, and for the rapid development of policies to assist appropriate social adjustments (see also Chapter 5).

1 *Post-industrialism.* Manufacturing has declined as a dominant employer, although its productive capacity is capable of further expansion unless anxiety about rates of resource depletion cuts this back. There has been a transition to a 'service' or post-industrial economy in which far more workers are employed in producing tangible and intangible services than in manufacturing goods.

2 *Costs.* The cost of technology has fallen dramatically relative to the cost of human labour. Despite inflation and the rising cost of resources, the price of each unit of performance in micro-technology is 100 000 times cheaper than it was in 1960. 'Miniaturization' has destroyed the historic relationship between the cost of labour and the cost of technology, permitting exponential growth with insignificant labour input. It is now possible (for the first time) to maximize two advantages – high outputs and low inputs – at once, which will lead to the reduction of labour in all high-volume process work.

3 *Range, capacity, reliability.* The new technology has far greater range, capacity and reliability than any which preceded it. Microchips increased their capacity 10 000 times between 1971 and 1980. Because microprocessors are programmable, they can readily be directed to stamp out metal, cut cloth, conduct psychotherapy or plan a school syllabus. The new technology is capable of working for twenty-four hours each day, performing many functions at once and generating little heat or waste. It is cheaper to replace modules in units than to repair them.

4 *Transitions.* Much precision equipment (e.g. cash registers, wrist-watches and telephone exchanges), the making and servicing of which required a large skilled labour force, has been replaced by purely electronic systems which require fewer workers.

Technology is moving from complexity towards simplicity (not the other way about, as has been the normal historical experience). Plastics, generally less labour- and energy-intensive to make and process, are replacing metals.

5 *Speed of diffusion.* Rates of development, adoption and dissemination of technologies devised in the 1960s have been unprecedentedly rapid. (The new technology of the First Industrial Revolution was comparatively slow in its adoption and diffusion: for example, the first steam-powered cotton mill in the United States dated from 1847 – sixty-three years after its adoption in Britain.[23] Low cost means that less developed countries (LDCs) can acquire the latest technology – and the combination of high technology and low wages makes them formidable competitors. Virtually every area of manufacturing, tangible services, or information-based employment will be profoundly influenced by new technology. Typically, each 'generation' of new technology (measured in years, not decades) is cheaper and more efficient than its predecessor.

6 *Speed of operation.* Computer transactions are now capable of being measured in multiples of picoseconds or 10^{-12} (1 000 000 000 000 picoseconds are equal to one clock second, or the same number of clock seconds as there are in 31 710 years). Information can now be disseminated and retrieved throughout the world in microseconds. The relationship between computers and telecommunications multiplies the power of both, and the capacity for instant, universal communication is unprecedented.

7 *International economy.* No economy (to re-apply John Donne) is an island, entire of itself. All advanced capitalist nations, and many in the communist blocs and the Third World, are now largely interdependent: a boom or a collapse in one area may have dramatic effects elsewhere. This reduces the influence of individual national economies and of specific local or regional markets; it also leads to the international division of labour, and the growth of power by multinational corporations such as Exxon, ITT, General Motors, IBM, Royal Dutch Shell, Ford, General Electric, Unilever, Philips and many others.

8 *'Smart machines'.* The new technology, for the first time in human history, does not merely extend or replace physical capacity but may also extend or replace human mental capacity. Artificial intelligence (AI) has been defined by Martin Minsky as 'the science of making machines do things that would require intelligence if they were done by men'. Computers can be programmed to parallel human mental processes – including the exercise of judgment and intuition. Machines have never before posed a challenge to mental (or white-collar) work: no steam engine could play chess, no slide rule could converse.

9 *Destructive capacity.* For the first time in history, man has the

technical capacity to destroy all life on this planet.[24] The extent of this potential catastrophe has no parallel. Technological determinism – the same secular religion which created the uranium, plutonium and hydrogen bombs – is a major factor in promoting the Western model of technological development (whether appropriate or not) throughout the world: this is one area where capitalists and communists are agreed.

10 *Decline of ideology.* The sheer complexity of new technology has contributed to a decline in the political process. Few politicians understand why and what is going on in technologically sensitive areas and thus tend to concentrate on peripheral matters: administration, dominated by bureaucrats and technicians, has displaced politics. (The implications of this phenomenon are examined in Chapter 8.)

11 *Population.* The First and Second Industrial Revolutions were marked by very rapid rates of population growth in technologically advanced countries. The provision of urban services – piped gas, reticulated water and sewerage, electricity, houses, roads and public transport – was a major form of employment. After 1970, life expectancy stabilized and birth rates fell in rich countries, although a population explosion continues in the Third World countries.

12 *Resources.* The First Industrial Revolution relied on abundant raw materials produced in relatively advanced economies, and the Second on cheap raw materials – rubber, oil, aluminium, copper – from less developed countries. After 1970 there was growing anxiety about the depletion of resources – the United States, for example, with 5 per cent of the world's population, was consuming 40 per cent per annum of its non-renewable raw materials. How many nations could, or should, follow this example?

13 *Transference.* In the past as some forms of employment (e.g. making horse-drawn vehicles) declined, new forms (e.g. making motor cars) grew and job transitions could be effected, often in the same areas. The development of an international division of labour means that work decline in one area may not lead to the growth of 'complementary' work: e.g. the decline of the conventional watch and clock industry in Switzerland and West Germany was followed by a rise in employment in making digital watches and clocks in Japan and South Korea – but there was no personal transition for the actual workers affected.

14 *Labour-displacing technology.* Sewing machines, typewriters, manually operated telephones and motor cars were all designed to have one operator for each machine and they *did* generate enormous employment – not only in operating but also in making, selling and maintaining. Much technological innovation in the past was 'labour-

complementing' – it extended the capacity of the existing labour force, and the machines themselves changed the nature of work. But there has been a significant shift to 'labour-displacing' technology where low-cost machines are specifically intended to reduce, if not eliminate, labour inputs. (The distinction between 'labour-complementing' and 'labour-displacing' technology is examined in Chapter 4.) The impact of this is particularly significant in information-based employment.

15 *Demand and employment.* The conventional wisdom that employment levels are essentially determined by demand is of diminishing relevance. Some employment is intensely demand-sensitive (e.g. personal services such as hairdressing, dentistry, dry-cleaning, beauty care), but in the massive supply of goods and general services employment appears to be in *inverse* relation to demand (see also Chapters 4 and 6).[25]

Few economists have addressed themselves to these issues. Their thinking has been shaped by the axioms of the Industrial Revolution, and they are unable to entertain the possibility of radical discontinuity with the past. Australian economists writing about current problems, for example, give no hint of recognizing that Australia has passed through a massive transition from a goods-based economy to a services-based economy. It is as if vulcanologists were asked to comment on Pompeii and Krakatoa, and replied 'What are they?'. Attempts to apply old remedies to a new situation are as futile as attempting to use a Ptolemaic world-view to interpret an Einsteinian universe.

Looking for the 'X' industry

As already outlined, the three industrial revolutions since 1780 have been marked by the development of major technologically based and labour-complementing industries, often appearing in clusters. The major developments (set out in the diagram on p. 00), from steam engines to the growth of telematics – some stimulated by wars or rumours of wars – have taken up the employment slack caused by technological advance.

Those who argue the case for continuity are confident that a huge new labour-absorbing industry – comparable to railways, cars and aircraft – will arise in the future. Boundless confidence that the mysterious 'X' industry will prevent massive unemployment is accompanied by a coy reticence in identifying it; the optimists are unable to point to its emergence anywhere in the world.

I believe that X will indeed arise, but it will not be as the conventional economists suppose. If X *does* appear, it will not be based on a new invention or technological form – rather it will involve an extension of labour- and time-absorbing work, the type of employment which does not face direct competition from technology, either new or old. Education is the area with the greatest potential in this regard, followed by home-based industries, leisure and tourism, skilled craftwork, and welfare services. Such an 'industry' will, however, depend on our developing a more co-operative (or *convivial*) form of society. Aggressive competition and welfare cost-cutting will, if pushed to their ultimate limit, destroy Western capitalist society. Economic growth depends on rising levels of income and an equitable spread of both income and employment. If technological advance leads to wholesale unemployment, destitution and further concentration of wealth in very few hands, then the whole rationale of a consumption-based capitalist economy would be destroyed. Successive industrial revolutions have led to increased spending power (although periods of slump have caused acute hardship to large minorities), and the new goods and services have been consumed. But can technology accelerate at such a rate that its product is not purchasable? If so, we would see the realization of Karl Marx's prophecy in Volume I of *Capital* that monopoly power would be concentrated in so few hands that capitalism would lose its profitability and 'the expropriators will be expropriated'. Inefficiency and waste could be important elements in X – we consume more than we need, and this assists employment. Many firms and administrative systems buy the most advanced hardware, and then employ it at reduced capacity or fail to understand how it works and what it could do (see also Chapter 5). Further, and perhaps not clearly perceived, is the fact that society may be reluctant to push the adoption of high technology to its ultimate limits for fear of creating personal or social disorientation.

The optimistic belief that the new X industry need not be thought about or planned for reveals a staggering ignorance of economic history. The great changes in technology between 1780 and 1914 – e.g. mechanization, the use of coal rather than charcoal and the replacement of wood by iron and steel, flying machines, new medical techniques, modern weaponry, photography, refrigeration, and the use of electricity for light and heating – had all been contemplated for generations, if not centuries. They did not spring up like mushrooms. New products and new technology are, after all, designed to fulfil human needs. It is doubtful if there are any absolutely *new* needs: new products are essentially only variations of age-old needs for food and drink, heat, shelter, transport, education and entertainment.[26] It is sometimes argued that new products or job opportunities will come out of the blue. Why? They never have.

In the thirteenth century, Roger Bacon wrote about horseless carriages, flying machines, powered boats, the use of spectacles, gunpowder, and the construction of automata. In *The Myth of the Machine* (1967) Lewis Mumford noted that Leonardo da Vinci 'had foreseen in his usual enigmatic way (labelled a dream) that "men shall walk without moving [motorcar], they shall speak with those absent [telephone], they shall hear those who do not speak [phonograph]".' He also experimented with flying machines. Johannes Kepler (1609) and Bishop John Wilkins (1638) both wrote at length about the technical problems of a flight to the moon. Precision engineering began in the seventeenth century with astronomical equipment, timepieces and automata. In the eighteenth century, Restif de la Bretonne predicted aerial flights over Australia. In 1804, Sir George Cayley produced a model flying machine which lacked only a lightweight power source to succeed. Charles Babbage began work on his model computer in 1834. In 1846, Herman Melville predicted that Americans would be flying to Hawaii for weekends. Writing in the 1860s and 1870s, Jules Verne gave very detailed descriptions of submarines, motion pictures, television, space travel and a variety of household devices. Edward Bellamy, in *Looking Backward* (1887), and H. G. Wells (writing from the 1890s), had precise ideas about what people would be working at in the late twentieth century.

The American historian Henry Adams (1838–1918) grasped intuitively that rates of energy use and acceleration were the most accurate indicators of technological change, and in 1906 wrote:

> It is quite sure, according to my score of ratios and curves that, at the accelerated rate of progression since 1600, it will not need another century or half century to turn thought upside down. Law in that case would disappear as theory or *a priori* principle and give place to force. Morality would become police. Explosives would reach cosmic force. Disintegration would overcome integration.

From the Wright Brothers at Kitty Hawk (1903) to Neil Armstrong on the moon was only a period of sixty-six years. From the first electronic computer (1946) to the micro-processor was only twenty-five years. From the first laser (1960) to its use on moon probes was less than a decade. The gap between discovery and ultimate form in inventions is closing. It is sheer intellectual laziness to say, 'We don't have to identify future areas for developing goods and services. Nobody did in the past.' In fact this *did* happen habitually in the past.[27] It is certainly true that new and hitherto unimagined techniques or components have been developed – e.g. silicon chips or memory bubbles – but nobody buys or makes chips or bubbles for their own sake, only as components in computers or other electronic devices.

We may also anticipate the development of new industries in the 1980s and 1990s – fusion for energy, biotechnology, fibre optics, printing at a distance, the use of 'smart machines' in retail distribution, communication satellites, electronic games, new sedatives and stimulants. But which of these will be labour-intensive? Which will displace employees in already existing industries? Biotechnology (splicing genes from recombinant DNA into plasmids from specific bacteria to 'manufacture' desired living materials) will be enormously productive and ubiquitous in the 1980s, but it will not be the 'X' industry. There will be net losses in employment in existing chemical, synthetics and pharmaceutical industries. Biologically created insulin will be far cheaper than insulin extracted from animals. Animal feed can be produced from oil waste processed by biotechnology, with fewer pollution problems and more grain available for human consumption or as an energy source. Micro-technologies will assist the crippled to propel themselves, the deaf to hear, the dumb to simulate speech. But is it suggested that any of these developments – or all combined – will create industries which employ as many people as motor manufacturing does now? For years, Australian economists put great faith in air conditioning as a future growth area.[28] Herman Kahn suggests space colonization; Albert Rosenfeld argues for services related to extreme longevity; cryonics (snap-freezing of the comatose in the hope that future medical advances will enable restoration to health after defrosting) has its advocates; the *Economist* is attracted to the computerized ski-boot, where danger signs will cause the boot to eject automatically from the ski. But will they be big employers? Much has been done to explore the universe's extremities, from quasars and black holes in outer space to sub-atomic particles such as quarks. But it is difficult to see where and how these extremities can be extended, let alone provide practical applications to meet human needs – or employment.

Is it suggested that human beings will develop an absolutely new range of functions which will require as-yet-undreamt-of technology? It is striking that when contemporary writers such as Isaac Asimov, Arthur C. Clarke and Kurt Vonnegut Jr write of future industry they describe it as capital-intensive, with very few workers.

Towards an age of asymptote?

The inventive peak of the Third Industrial Revolution was between 1942 and 1970, coinciding with the era of full employment. The 1970s marked a significant decline in major discoveries, although many new technological refinements were produced. In aviation, for example, the jumbo jet dates from 1968 and the supersonic Concorde from 1969: the wide-bodied jets of the 1970s were not innovative. In micro-electronics, large-scale integrated circuits date from 1969 and the micro-processor from 1971. Magnetic bubble memories were developed in 1966 and the 'Josephson junction' in 1968: since 1970 their capacity has increased enormously, but again few new concepts have emerged (see Chapter 5). Stephen Hawking's brilliant work on 'black holes', published in 1971 and extended in 1974, seems unlikely to become the basis of a major industry. Orbiting space laboratories and communications satellites were put into service in the 1970s based on techniques first used during the space race nearly twenty years earlier. Lasers and radio telescopes were products of the 1960s.

Has technological development reached *asymptote* (here used in the general sense of the flattening-out of a curve after a long exponential rise) – a ceiling limit? In some areas this seems possible: we live in an era of instant communication, where messages flash round the globe in one-seventh of a second; experimental computer operations are now being measured in picoseconds, and computer scientists talk about 'real time' (i.e. simultaneous operation). How do we improve on that? Miracles are being achieved, at enormous cost, in microsurgery. Where do we go after that? Existing weapon capacity could cause 4500 million deaths. Who could want for more?

Human capacity is limited by one inflexible factor – time. For decades we have had the technological capacity to conduct individual conversations with tens of thousands of people every day, but this is meaningless in practice if we can only devote a fraction of a second to each of them. There are physical limitations on speed: moving at more than 25 600 kilometres per hour, a vehicle would be thrown out of the earth's orbit. High speeds may be psychologically disorientating and economically pointless, especially when we can send words and images at the speed of light with infinitely lower cost. The average speed of vehicles in London or New York streets is less in 1980 than it was in 1910, so faster mass-produced vehicles seem an unlikely prospect. Future developments may be more likely in *lateral* extension – spreading existing or modified technology more widely throughout the world – rather than a *linear* extension into new forms of specialized technology.

Conclusion

A careful analysis of economic history suggests Jones' First Law:

Technological innovation tends to reduce aggregate employment in the large-scale production of goods and services, relative to total market size, after reaching maturation (e.g. transition from manual telephone systems which were labour-complementing to automatic systems which were labour-displacing), and to increase employment at lower wage rates in areas complementary to those technologically affected.

The mechanization of agriculture led to the textile mills of Bradford and Leeds. Whitney's cotton gin intensified and prolonged slavery in the USA. The steam-driven textile mills enabled Karl Marx to have a live-in servant. The computer revolution indirectly created jobs for dish-washers at McDonalds. I adopt the views of J. S. Mill, Karl Marx, Michal Kalecki, J. M. Keynes, Lewis Mumford, Simon Nora and Wassily Leontieff that the continued reduction of necessary labour inputs will lead to chronic unemployment, especially for the unskilled unless new forms of work, *complementary* to technology, are encouraged.

The 1970s were not only a period of world-wide economic downturn, but were also notably deficient in intellectual, artistic and scientific achievement. I am not arguing that poor-quality novels, films or paintings (for example) are caused by a trading slump – or vice versa – but periods of economic growth appear to coincide with intellectual and artistic vitality.[29] The decade was marked by an exhaustion of intellectual curiosity, growing pessimism about education, spiritual apathy and social anomie, a rise in irrational cultism, a retreat to privatized experience and drug dependence, a loss of historical perspective (with increasing emphasis on 'now time'), dropping out and alienation. It is unreasonable to look for steady, straight-line growth from the 1960s onwards. I agree with the American economic critic Hazel Henderson (see Chapter 4) and the Canadian GAMMA group (see Chapter 10) that the most appropriate analogies for economic processes are to be found in biology – with growth, maturation, nourishment, excretion and decline – rather than in physics.

While Kondratiev's 'long waves' do not coincide absolutely with the periods of intense intellectual and economic activity in what I have called the First, Second and Third Industrial Revolutions, it seems likely that 1970 marked the end of an upswing and the beginning of a downswing which may continue throughout the 1980s. The low birth rates in advanced countries also suggest a decrease in confidence. The decline of the political process has been particularly marked in the 1970s and 1980s. While Jimmy Carter and Ronald Reagan are the political heirs of Thomas Jefferson and Abraham Lincoln, it is difficult to believe that they represent an improvement in quality – despite universal literacy, an omnipresent media and a vast information industry. The fragmentation of knowledge, increasing personal dependence on technology, a decline in social relationships, the atomizing of society, and spiritual exhaustion due to trivialization appear to constitute major threats to human personality. A pre-occupation with materialism, a conviction that national and international salvation is to be found in economic growth alone, and emphasis on externalized (consumption-based) value systems confirms Jacques Ellul's observation that 'there has been a disappearance of political ideologies and a proliferation of substitution ideologies . . . Increasing material power goes hand in hand with decreasing ideological power.'[30] Concern about the rate of depletion of the world's resources, especially energy, may encourage the revival of decentralized labour-intensive forms of work, but it may also encourage dependence on micro-electronics – which is labour-displacing and parsimonious in the use of energy and space.

I conclude that the 'post-industrial revolution' is, like the First Industrial Revolution of the 1780s, a fundamental break with previous economic history. But even if we accept that the present era represents full continuity with the past, on the basis of the historical record we would have to assume that long-term instability in employment will be the norm.

3 A New Analysis of the Labour Force

The trouble isn't what people don't know; it's what they do know that isn't so.

Will Rogers

The three-sector analysis of the labour force conventionally used by most British and Australian economists, journalists, public servants and politicians roughly equates primary industry with growing and digging things up, secondary with making things, and tertiary with doing things. This was first proposed by the New Zealand economist A. G. B. Fisher in *The Clash of Progress and Security* (1935), where he applied the term 'tertiary industries' to employment 'not covered by the titles current in Australia and New Zealand of "primary industry" for agriculture, grazing, trapping, forestry, fishing and mining, and "secondary industry" for manufacture'.[1] Fisher's analysis was refined and popularized by Colin Clark in *The Conditions of Economic Progress* (1940).

1 *Agriculture:* 'the common feature . . . is [dependence] upon the direct and immediate utilisation of natural resources . . . [which] can only be carried out at the point where the natural resources are . . . Mining is a border-line case, which is sometimes included here, sometimes with manufacture, and which perhaps deserves a class to itself.'

2 *Manufacture:* 'a process, not using the resources of nature directly, producing, on a large scale and by a continuous process, transportable goods. This definition excludes the production of untransportable goods (buildings and public works), and small-scale and discontinuous processes such as the hand tailoring of clothes, or shoe-repairing.'

3 *Service industries:* 'These naturally group themselves further into building and construction; transport and communications; commerce and finance; professional services; public administration and defence; and personal services, of which private domestic service may be distinguished from commercially supplied services, such as cafes and hairdressing.'[2]

The rapid increase in 'tertiary' or 'service' employment after World War II, and its growth to between two-thirds and three-quarters of the paid labour force in sophisticated economies, makes the Fisher-Clark analysis an anachronism. 'Services' is now far too broad and amorphous a category – it does not take into account the changing nature of work, which is occurring most within the tertiary sector. The three-sector analysis does not distinguish between the motor mechanic, the computer programmer and the geriatric nurse, all of whom are covered by the 'services' umbrella; it fails to recognize the significance of information-based employment, now large and important enough to warrant a separate sector; it rejects the economic and social significance of home-based employment, especially of women; it ignores the growing 'informal' (cash) economy where work is not measured by wages paid and taxes collected – in hobbies, building and other DIY activities – and the development of a 'counter-economy' (or 'shadow economy'). To establish more precisely what people are actually doing, and to understand the significance of historic trends and shifts in employment, we must abandon the conventional three-sector analysis and divide the tertiary sector into two or three smaller elements. (This has also been attempted by Nelson Foott and Paul Hatt, Jean Gottman, Daniel Bell and others.)

As a first step, I will introduce a four-sector analysis of the paid labour force – placing information in a separate category – proposed by the American economists Dr Marc Uri Porat and Professor Edwin B. Parker.

The four-sector analysis

The Porat-Parker analysis, proposed at an OECD conference on computer telecommunications policy held in Paris (February 1975), has been widely accepted in the USA and to some extent in Australia:

Primary: extractive.
Secondary: manufacturing and construction.
Tertiary: services not based on the transfer of information.
Quaternary: information (including printing).

The *primary sector* covers agriculture, forestry, fishing, mining, quarrying and oil extraction. The *secondary sector* involves 'making things': using skills and/or techniques on processed or raw materials to produce finished goods. Building and construction are treated as quasi-manufacturing processes and removed from 'services' where they are generally found. Printing (but not papermaking or bookbinding) is removed from this sector because its essence is the transmission of information rather than making something for its own sake.[3] The *tertiary sector* provides 'hard', or tangible, economic services, involving the processing or transfer of matter and/or energy. These services are easily quantifiable and have a precise economic value to consumers. Tertiary activities include transport, storage, buying and selling goods, water and energy supply, maintenance, waste disposal, cleaning, beauty care, heating and cooling, supply of food and drink, sports and recreation, and many services performed by armed forces, police, doctors and dentists. In this sector labour is likely to remain 'demand-sensitive', because most of the services depend on personal participation at specific locations (e.g. supplying petrol to a million cars involves a million separate transactions). Direct physical involvement is central to tertiary-sector transactions, although some reticulated utilities (water, gas, electricity) have a diminishing labour input.

The *quaternary* (information) *sector* provides 'soft', or intangible, services and its common element is the processing of symbols and/or symbolic objects. Its services are often difficult to quantify, and their value to consumers is often more subjective than objective. Information activities include teaching, research, office work, public service, all forms of communication and the media, films, theatre, photography, posts and telecommunications, book publishing, printing, banking, insurance, real estate, administration, museums and libraries, creative arts, architecture, designing, music, data processing, computer software, selling tickets, accountancy, law, psychiatry, psychology, social work, management, advertising, the church, science, trades unions and parliaments. The symbols used are words, sounds, images and numbers, or symbolic objects which represent tangibles (e.g. money, cheques, letters, speeches, bills, tickets, type, photographs, advertisements, keys, title deeds, legislation or newspapers). The tools of trade used by information workers are voices, pens, pencils, chalk, telephones, typewriters, computer keyboards, word processors, duplicators, visual display units (VDUs), adding machines, microphones, cameras, microscopes and tape recorders. Information-based employment increased rapidly until the 1970s and the sector has been very labour-intensive. However, computer-based technology will reduce 'soft' employment which is not dependent on physical presence or particular locations to effect transactions (e.g. airline bookings, banking, typesetting for newspapers, and data processing).

INFORMATION EMPLOYMENT IN THE UNITED STATES

In 1977, the US Department of Commerce published Porat's *The Information Economy*, a nine-volume computerized analysis of information employment in the United States. Porat defines information as 'data that have been organised and communicated. The information *activity* includes all the resources consumed in producing, processing and distributing information goods and services.' In examining census returns, Porat asked the following question (and considered its implications):

> Does this worker's income originate *primarily* in the manipulation of symbols and information? Clearly, all human endeavour contains some component of information processing. Without information processing, all cognitive functions would cease and there would be no human activity. But that definition is operationally useless. We are not saying that information workers deal exclusively in information and other kinds of workers never deal in information. Rather, we assert that certain occupations are primarily engaged in the manipulation of symbols either at a high intellectual content (such as the production of new knowledge) or at a more routine level (such as feeding computer cards into a card reader). And for other occupations, such as in personal service or manufacturing, information handling appears only in an ancillary fashion. It is a distinction of degree, not of kind.

The graph on p. 50 shows the relative distribution of the US labour force, using the four-sector analysis, since 1860. There was a steady growth in 'information' employment from 1860 to 1940 (between 1880 and 1890, for example, the numbers grew from 1·1 million to 2·8 million, an increase of 15 per cent per annum). Growth then became rapid until 1970 – by 1967, 53 per cent of all income was paid to information workers. However, between 1970 and 1980, the proportion of information workers increased only fractionally (46·4 per cent to 46·6 per cent) although there was a modest increase in absolute numbers in a decade of very high employment growth, with the greatest rises being in the supply of physical services (transport, food and drink, health, leisure and recreation).[4] Porat argues that the simplest and most dramatic illustration of the information economy would be to adopt a two-sector analysis of the labour force, divided into 'information workers' and 'non-information workers'.

What are the prospects for continued employment growth in this sector, already 30 per cent of the paid labour force in Australia but which declined slightly between 1976 and 1979? Much information processing will be done on an individual basis and will remain labour-intensive, e.g. social work and counselling, much secretarial and office work, research and development. Group work – such as the relation-

Four-sector aggregation of the US labour force, 1860–1980
(Using median estimates of information workers)

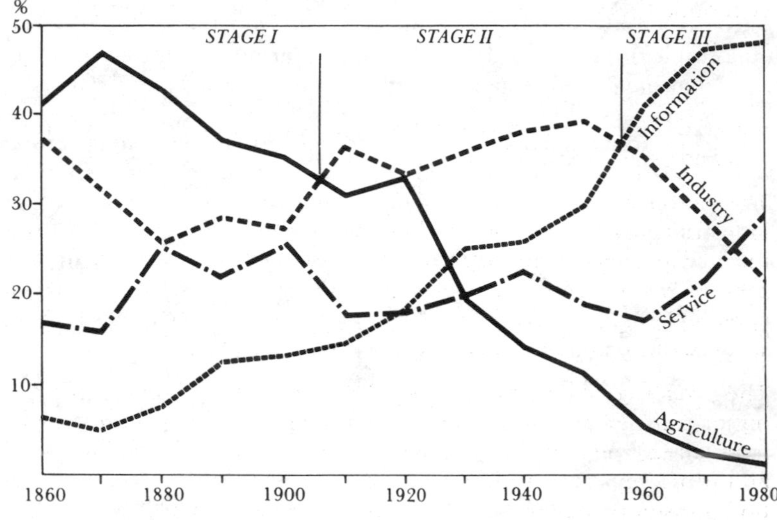

Source: Porat, vol. I, p. 121.

ship of the teacher with a class, or of a manager with a small labour force – will also continue to be extremely important. There is, however, a problem in the area of mass information services, where computers are replacing armies of clerical workers in the public service, banking and insurance, and in the dissemination of information through print and electronic media. While there is a vast, sometimes exponential, increase in the *volume* of transactions, labour intensity is being replaced by capital intensity due to technological advance.

The information sector is far more susceptible to the 'bite' of new technology – in terms of cutting costs and increasing productivity – than are agriculture and manufacturing, and it seems inevitable that employment in this sector will contract in the 1980s. It is essential that Australian economists, statisticians and politicians recognize the significance of the information sector in making labour-force projections. Nevertheless, the four-sector analysis is limited to paid work and is now an inadequate tool for understanding the changing nature of employment.

The five-sector analysis

I propose a five-sector occupation analysis of an extended labour force which includes men and women working without remuneration within the home, unpaid community work, and home-based economic activity such as craftwork and DIY.[5] The groupings are as follows:

The *primary sector* (extractive) comprises the production of basic materials, agriculture, forestry, fishing, mining, quarrying, oil extraction and growing crops for energy (biomass).

The *secondary sector* (manufacturing and construction) includes all industrial processes by which basic materials are converted into a finished product. It includes not only the manufacturing of motor cars, television sets, carpets, refrigerators, tyres, ice-cream, breakfast cereals, beer and cigarettes, but also the construction of houses, schools, office buildings, bridges and roads.

The *tertiary sector* (tangible economic services) is based on the processing or transfer of matter and/or energy. It includes moving people, maintaining their bodies and hardware, heating and cooling them, and providing them with goods and services not based on information transfer or analogous to home-based employment.

The *quaternary sector* (information processing) involves the processing of symbols and/or symbolic objects as already set out in the four-sector analysis above. The central role of symbols or symbolic objects is illustrated in the work of bankers or real-estate agents, where the transactions essentially consist of transfers of information, the negotiation of cheques, the granting of overdrafts, persuading clients, and transfers of names on to title deeds.

The *quinary sector* (domestic and quasi-domestic servicing and/or making) has as its common elements the provision of domestic services (generally unpaid), professional services analogous to domestic work, charitable work and hobby-based occupations. It may be subdivided as follows:

(a) *Unpaid work.* Care of families, especially children and the aged, home nursing, food preparation, house care and maintenance, washing and cleaning. Voluntary work to assist community-based welfare organizations which provide services originally provided in the home. Substituted professional services: home renovations and construction, gardening, boat building, domestic dressmaking and other DIY activities.

(b) *Home-based work where remuneration is incidental.* Craft-based activity carried out at home, such as pottery, enamelling, jewellery;

hobby-based activities such as producing flowers, vegetables or eggs, or the collecting of (and trading in) paintings, books and stamps which may produce incidental economic gain but where the activity is primarily undertaken for its own sake – i.e. *where work is itself an output of production.*

(c) *Professional provision of quasi-domestic services.* Food, drink and shelter: hotels, motels, restaurants. Care of children and the aged (but excluding educational services): nurseries, creches, old peoples' homes, day-care services for invalids. Gardening, lawn-mowing, the care of pets. House-cleaning, laundry and dry-cleaning, minor home repairs and maintenance. Miscellaneous home comforts: but also massage parlours.

My five-sector analysis *follows* the four-sector analysis by: (i) removing building and construction from 'services' and adding them to 'manufacturing' in the secondary sector; (ii) dividing 'services' into 'information' and 'non-information'; and (iii) treating printing as a medium, removing it from manufacturing and adding it to the 'information' (quaternary) sector. It *differs* by: (i) recognizing home-based employment, whether paid or unpaid, as a fifth sector; and (ii) removing professional services which are analogous to home-based employment – e.g. the care of children and the aged, cleaning, catering and accommodation – from the tertiary sector and adding them to the quinary. As Australia moves from a post-industrial to a post-service society, it may be necessary to adopt a sixth (senary) sector for students – who, like housewives, are not classified as workers.

ADVANTAGES OF THE FIVE-SECTOR ANALYSIS

The five-sector analysis has the following advantages:

1 It provides a maximum of information with a minimum of complexity, and would be a valuable tool for future labour-force projections.

2 It acknowledges that housewives, for example, or retired people active in community work, are entitled to be recognized as workers and eliminates some mythology about the composition of the labour force.

3 It makes it easier to recognize which areas of employment are growing and contracting.

4 It will assist in recognizing the apparent worldwide tendency towards low-volume high-employment activity as industrially based high-volume low-employment opportunities contract.

5 It places employment in a context of time use and life commitment rather than income. For example, in current labour-force enumerations, if a housewife works for one half day each week in a greengrocer's shop she is recorded as if she were a full-time shop assistant and her domestic work has no statistical recognition.

6 It recognizes the economic and social significance of the 'informal economy', which is ignored by statisticians.

7 It points to the inadequacy of current methods of measuring production. In *Home, Inc.*, Scott Burns estimates that if the US gross national product (GNP) included the value of households and work performed in them, the total would equal the entire amount paid in wages by every American corporation.[6] He also notes that statisticians only value the household when it breaks down – i.e. when substitute welfare services have to be provided.

8 It emphasizes the economics of *use* value rather than *market* value – and this suggests that households might be granted tax deductions for capital items used in the course of social production in the same way as corporations.

9 It notes the relationship between paid and unpaid work which is similar in type (e.g. care of the aged, gardening).

SOME PROBLEMS OF CLASSIFICATION

My five-sector analysis provides vertical divisions, but it does not recognize horizontal classifications of the labour force or society – i.e. class or hierarchical divisions. Other descriptive terms should be included to distinguish between full-time and part-time employment, public and private employment, and those who are self-employed or employed.

The five-sector analysis should be read in conjunction with the Australian Standard Industrial Classification (ASIC), published by the Australian Bureau of Statistics to conform with the International Standard Industrial Classification of All Economic Activities (ISIC) prepared by the UN Statistical Office. ASIC is a fine-grained index of industrial activity, beginning with thirteen divisions:

A: Agriculture, forestry, fishing and hunting.
B: Mining.
C: Manufacturing.
D: Electricity, gas and water.
E: Construction.
F: Wholesale and retail trade.

G: Transport and storage.
H: Communication.
I: Finance, property and business services.
J: Public administration and defence.
K: Community services.
L: Recreation, personal and other services.
M: Non-classifiable economic units.

These divisions are then broken down into subdivisions and groups, each given its own number code, so that thousands of industries can be readily identified. This index does not, however, assist in identifying structural change in *types* of work. 'Motor manufacturing industry', for example, might include a hundred or more types of work – and yet in the ASIC classification typists, tradesmen, cleaners and managers are placed together. A newspaper report that 1200 jobs were lost in the motor industry may be misleading unless we know what type of jobs were lost – process workers or office workers, skilled or unskilled. In education, teachers, office workers, cleaners and librarians work in the one 'industry' but their functions and scope for job transfer are quite different.

My five-sector analysis of the labour force can be illustrated by a homely example – the bread industry – which suggests the interdependence and relative size of each sector. This could be seen as an inverted pyramid:

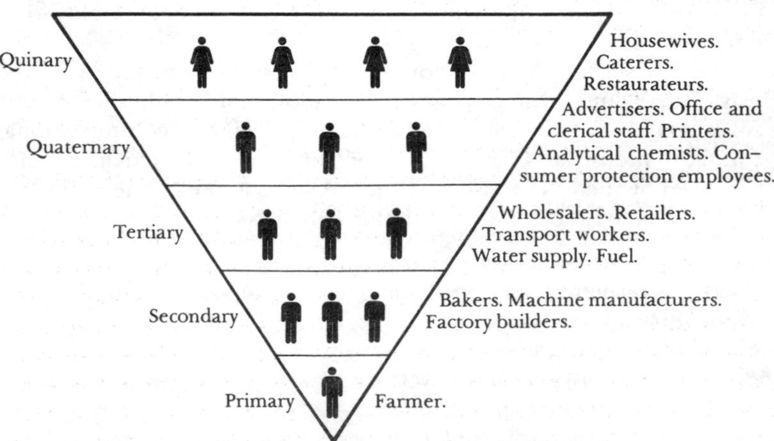

Of the fourteen people involved in this example, one is producing raw material (wheat) and only four (primary and secondary sectors) are personally involved in making any tangible product. The others move the wheat to and from the bakery to point of sale, advertise it, sell it, pick it up and use it. The pyramid helps to explain both why farmers believe their work is not adequately rewarded (although the sense of proprietorship may provide psychological consolation), and the huge overheads which make bread prices very high.

A further problem is that some areas of employment are difficult to classify – defence, police and health care. Police and defence personnel perform a variety of roles which may be classified as tertiary, quaternary or quinary where appropriate. When a doctor gives advice he is in the information (quaternary) sector; when pulling out stitches, he is in the tertiary sector. (While delivering babies he may be in the primary – extractive – sector.)

Finally, the basic weakness of any type of labour-force analysis is its dependence on self-description in census returns or other surveys, which may give a misleading impression of what people are actually doing. For example, when a housewife is working sixty hours per week running a home and twenty hours per week as a librarian, the census return in effect forces her to record only one occupation: her (arbitrary) choice will decide whether the quaternary or quinary sectors rise or fall by one. A teacher working as an army officer selects his own category – if he describes himself as a 'teacher' or 'officer', the returns are varied accordingly. Furthermore, the census does not, of course, record multiple job-holding or 'moonlighting'. Such limitations, however, probably cancel themselves out so that the figures provide a reasonably satisfactory picture of long-term trends.

Labour-force analyses applied to Australia

Australia was probably the first 'service-based' economy anywhere. Unlike Britain, the United States and most of Western Europe – where industry was dominant for decades after the decline of agriculture and before 'services' became pre-eminent – manufacturing has at no time been the largest employer in Australia.

Services displaced agriculture to become the largest employer in this country some time between the colonial censuses of 1861 and 1871, shortly before Australia began its own limited 'industrial revolution' with technology and investment capital imported from Britain. Our unusual pattern of settlement and population increase was responsible for this early growth of 'service' employment. From 1788, population was concentrated in a few urban centres surrounded by huge tracts of open country. The five mainland colonies developed along similar lines – the metropolis was eight to ten times more populous than the second town, and the capitals were long distances apart. This urban pattern remained: with almost 56 per cent of its population in five large cities, Australia ranks with Japan as the world's most urbanized nation.

After convict transportation ended there were two decades of free immigration and settlement, followed by the freak conditions of the Gold Rush when Australia's population increased 166 per cent between 1851 and 1861. This multiplied the demand for commodities which people were not able to provide for themselves: manufactured goods were imported, and paid for by gold. Australia had its own railway mania in the 1860s. The capitals had been linked by telegraph by 1858, electricity supply and the telephone became increasingly common in the 1880s. The economic boom in that decade was based on wool, wheat, metals and house- and town-building – not on manufacturing.

With its strong resource base and rapid population growth, Australia almost certainly had the world's highest per capita income in the period 1870–90.[7] (By 1975 it ranked eighth, and in 1980 it was sixteenth.)

In the following table, a comparison of various countries with the United States (= 100) indicates the changes in GDP per worker-hour over the period 1870–1976. (In absolute terms, of course, the output of each US worker increased enormously in this period.)

Index of real GDP per worker-hour in selected countries, 1870–1976

	1870	1900	1950	1976
Australia	182	111	70	78
France	62	55	41	77
Germany, Federal Republic	63	62	35	81
Italy	60	39	31	64
Japan	24	22	14	51
Sweden	45	45	57	82
United Kingdom	122	94	55	62
United States	*100*	*100*	*100*	*100*

In the 1880s, for example, Melbourne had that proliferation of service employment which now characterizes sophisticated twentieth-century cities, with a university, museum, libraries, theatres, public transport and a great range of employment in agencies.[8] (Geoffrey Blainey says that Swanston Street had eighteen oyster bars.)

Australian statistics before the 1911 census are not very satisfactory because methods adopted by colonial statists differed so widely. Henry Heylyn Hayter, Victorian Government Statist 1874–95, attacked the New South Wales census for 1881 as an anachronism which ignored the significance of the railways, the telegraph and other innovations. 'No attempt is made to show the numbers engaged in manufacturing pursuits . . . The mode of grouping is such as might perhaps have answered sufficiently well 40 or 50 years since, but is quite out of date at the present time.'[9] Estimates of the relative size of labour-force sectors before this time should thus be viewed with scepticism and caution, but they do give some idea of employment trends. The 1881 Victorian census returns suggest the following breakdown of the paid labour force:

	Persons employed	*Percentage of labour force*
Agriculture	131 869	34·80*
Mining	38 764	10·23
Manufacturing	45 000	11·87
Services	163 288	43·10

* This is an increase on the 1871 Victorian census, where agriculture and fisheries accounted for 90 664 of a labour force of 294 181 (30·82 per cent).

These sectoral percentages are not necessarily representative of the other colonies, except perhaps New South Wales, since Victoria was the richest and most populous. Nevertheless, figures for agricultural employment in census returns for other colonies were fairly close to the Victorian figure: Queensland, 34 per cent; South Australia, 31 per cent; Tasmania, 38·5 per cent; Western Australia, 39 per cent.

On the basis of the four-sector analysis – with agriculture and mining in the primary sector, construction in with manufacturing (secondary), and services divided into tangible (tertiary) and information (quaternary – including printing), the 1881 figures for Victoria become as follows:

	Persons employed	*Percentage of labour force*
Primary	170 633	45·03
Secondary	65 648	17·32
Tertiary	121 766	32·13
Quaternary	20 874	5·52

Applying my five-sector analysis, which includes women performing unpaid home duties as part of an extended labour force, the following figures result:

	Persons employed	*Percentage of labour force*
Primary	170 633	35·55
Secondary	65 648	13·67
Tertiary	83 057	17·30
Quaternary	20 874	4·35
Quinary	139 744	29·11

The censuses for 1891 and 1901 were conducted on a reasonably uniform basis in all six colonies, and the first Commonwealth census took place in 1911.

The graphs and tables on pp. 60–61 cover the period 1891–1979.[10] For both the four- and five-sector analyses, the table provides figures in absolute numbers, and the accompanying graph illustrates the statistics as a percentage. The figures suggest slow and consistent changes in sectoral employment in Australia. (Employment shifts and factors relating to unemployment are discussed further in Chapter 6.)

EMPLOYMENT CHANGES IN THE SECTORS

Primary

This sector is overwhelmingly dominated by male employment. Farming has fallen steadily from less than one-third of Australia's paid labour force in 1891 to 6·8 per cent in 1979 in a four-sector analysis (10·5 per cent of males), or 5·2 per cent in a five-sector analysis. This is a stable employment area: while some disenchanted young leave cities to take up rural life, this is balanced by long-term movement of rural young to the cities. Australia has followed the secular changes in agriculture in Great Britain and the United States, where employment has fallen from over 50 per cent to about 3 per cent, with greatly increased productivity. India and China, on the other hand, with over 65 per cent of the active population in farming, are struggling to feed their populations.

Despite the minerals boom since the 1960s – including oil drilling and natural-gas production, where value of product increased from $244 million in 1960 to an estimated $3430 million in 1978 – mining has been extraordinarily stable as an employer: 0·7 per cent of the paid labour force in 1964, and 0·6 per cent in 1978. It is true that mining has a 'multiplier effect' in generating related employment – jobs in infrastructures such as mining towns, railway lines, port facilities, roads, heavy transport generally, and financial journalism – but it is hard to estimate the precise size of this effect, and figures hurled about are multipliers from six to fifteen. While the value and volume of its output will grow, mining will not increase as an employer. (In absolute terms there have been increases of a few thousand workers, but relatively there has been a persistent decline.) When new areas are developed, this will be done by a few skilled workers using sophisticated technology, not by vast armies of unskilled workers with spades.

There is growing concern that the strength of the mineral sector may have an adverse effect on employment in manufacturing. Dr R. G. Gregory of the Australian National University has argued that the growth of Australian mineral exports increases pressure on manufacturing by bringing about a *de facto* revaluation of the currency because it makes the balance of payments so lopsided.[11] The purchasers of Australian mineral exports want to pay, in effect, with finished goods which can be landed in Australia (discounting tariff protection) at a far cheaper price than they could be made for here. Gregory contends that mineral exports have been a greater factor for structural change than the 25 per cent general reduction in Australian tariffs in 1974. If tariffs were cut again, this would have a destructive effect on labour-intensive industries such as clothing, footwear and textiles. If the Australian dollar was officially revalued, it would make it harder for export-oriented industries to penetrate foreign markets. If tariffs are not cut and the currency is not revalued, then a resources boom is likely to generate high inflation. In the long term, increased mineral dependence will cause a rapid growth in money supply – but without a more equitable distribution of national wealth the gap between the rich and poor will grow, and there will be increasing pressure to convert Australian manufacturing to higher volume and lower employment. It may also increase Australia's role as an exporter of capital (and there is much evidence of this as Australian firms move 'offshore' into South-East Asia).

Secondary (manufacturing and construction)

The 1880s were a period of rapid growth in manufacturing, and by 1891 this sector (conventionally defined – excluding construction but including printing) comprised 20·59 per cent of the paid labour force throughout Australia. There has been a remarkable consistency in employment figures for manufacturing since that time – 20·4 per cent in 1911 and 21·9 percent in 1979 – apart from the period between World War II and 1971, which now appears anomalous. This consistency confirms that as old forms of work phased out (e.g. making

Four-sector analysis of the Australian labour force, 1891–1979 (%)

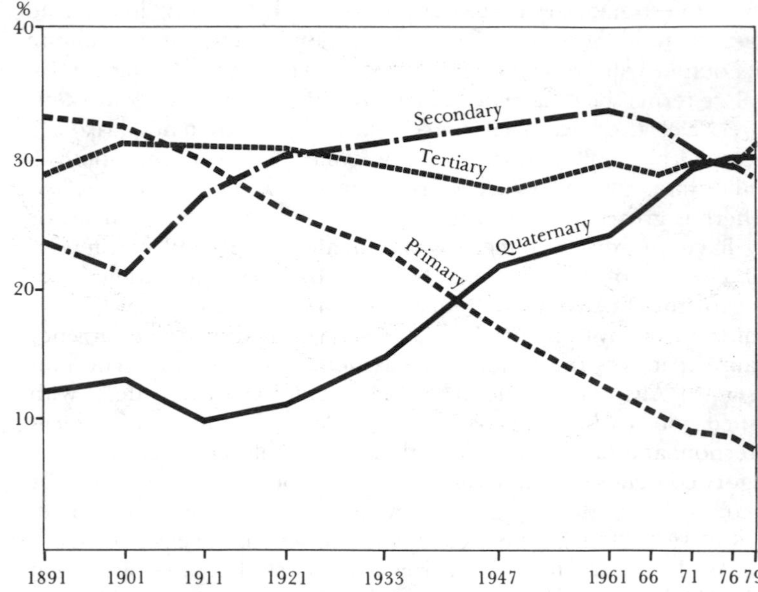

Note: Compiled at request by the Statistics Group of the Legislative Research Service from censuses of the Commonwealth of Australia and *The Labour Force* (May 1979), published by the Australian Bureau of Statistics.

Australian labour force classified by occupation, 1891–1979
(Four–sector analysis)

Year	Primary ('000)	Secondary* ('000)	Tertiary ('000)	Quaternary ('000)	Total ('000)
1891	440·8	321·9	382·9	176·4	1322·2
1901	527·8	346·8	509·9	231·6	1616·2
1911	586·1	541·0	609·9	197·7	1934·7
1921	598·7	698·4	722·4	266·5	2286·1
1933	647·9	855·8	827·1	407·4	2738·1
1947	547·2	1013·5	891·6	721·7	3174·0
1961	509·9	1398·2	1244·5	1024·9	4177·5
1966	512·8	1569·9	1406·4	1305·9	4795·0
1971	478·3	1616·4	1607·1	1563·5	5265·3
1976	514·7	1809·1	1820·4	1849·5	5993·8
1979	493·4	1856·4	2102·8	1986·1	6439·9

*Building and construction are included in the secondary sector.

Five-sector analysis of the Australian labour force, 1891–1979 (%)

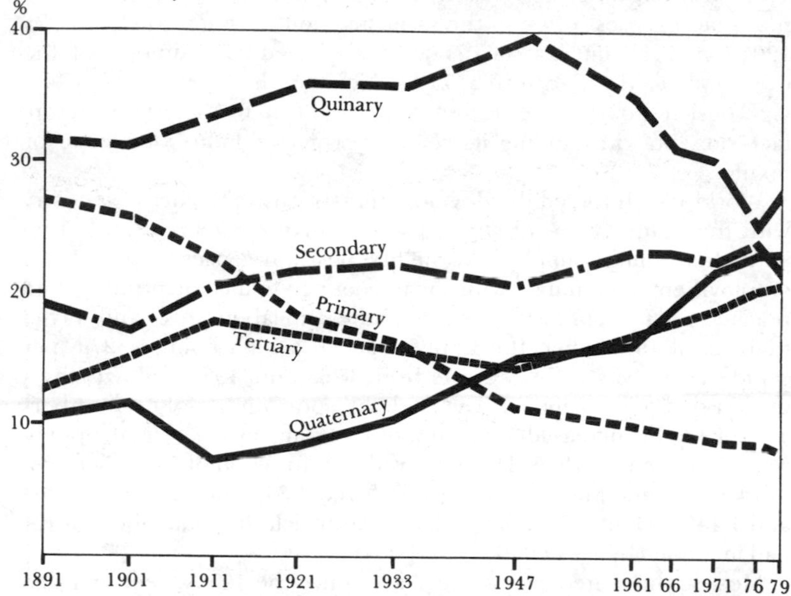

Note: Compiled at request by the Statistics Group of the Legislative Research Service from censuses of the Commonwealth of Australia and *The Labour Force* (May 1979), published by the Australian Bureau of Statistics.

Australian labour force classified by occupation, 1891–1979 (Five-sector analysis)

Year	Primary ('000)	Secondary ('000)	Tertiary ('000)	Quaternary ('000)	Quinary ('000)	Total ('000)
1891	440·8	321·9	223·0	176·4	549·5	1711·7
1901	527·8	346·8	308·8	231·6	658·2	2073·4
1911	586·1	541·0	437·2	197·7	893·9	2655·9
1921	598·7	698·4	540·9	266·5	1213·0	3317·5
1933	647·9	855·8	609·4	407·4	1447·0	3967·4
1947	547·2	1013·5	724·5	721·7	1949·3	4956·2
1961	509·9	1398·2	1047·3	1024·9	2160·2	6140·5
1966	512·8	1569·9	1172·4	1305·9	2126·2	6687·2
1971	478·3	1616·4	1354·6	1563·5	2216·4	7229·2
1976	514·7	1809·1	1598·6	1849·5	1970·3	7742·3
1979*	493·4	1856·4	1778·9	1986·1	2340·0	8475·8

* 1979 figures derived from a labour-force survey by the Australian Bureau of Statistics.

accessories for horse-drawn vehicles) new forms evolved (e.g. making motor accessories), often in the same area and with few job losses. By 1901, when all internal tariffs were abolished (Constitution of the Commonwealth of Australia, s. 92), New South Wales – which had vigorously espoused free trade – had caught up to Victoria in manufacturing (notwithstanding its years of protected industry) except for textiles.[12]

World War II forced the development of Australian heavy industry with the manufacture of ships, planes, tanks, artillery, ammunition and other military supplies. At the first post-war census, in June 1947, employment in manufacturing (again, defined to include printing) had reached a plateau of 25·05 per cent of the paid labour force and stayed there until 1965 when the figure reached a high point of 27·6 per cent.[13] There was a slow decline from 1965 until 1971, followed by a more rapid fall except for a very slight upturn in absolute numbers (stimulated by increased government spending) in 1973–74, dropping to 21·7 per cent in 1976. This involved a contraction of 21·38 per cent in twelve years. Manufacturing employed 1 312 000 people in 1966 and 1 147 000 in 1977, after a decade in which the total labour force had increased by 1 230 000.

Most sophisticated economies have, since the 1950s, been marked by an enormous growth in the output of manufacturing industry, accompanied by a steady fall in employment in that sector:

US	1950: 33%	1970: 29%	1980: 22% (est.)
France	1950: 45%	1970: 39%	1976: 28%
W. Germany	1950: 48%	1968: 48%	1976: 35%
UK	1951: 39%	1965: 35%	1976: 31%
Japan	1960: 28%	1970: 28%	1976: 26%

The size of the manufacturing labour force in Australia was kept artificially high in the post-war decades by a bi-partisan policy on tariff protection and, for some years, import licensing which encouraged Australian industry to concentrate on 'import substitution'. Manufacturing was limited by a small domestic market, and aided by an 'insular' economy where foreign competition could be easily excluded or ignored. Trade unions supported protection as an article of faith because it appeared to preserve jobs; Australian management lived in a small, sheltered world and a tariff wall was comforting to lean against. The fall in manufacturing employment would have occurred earlier but for the policies vigorously promoted between 1949 and 1971 by Sir John McEwen in a variety of ministries relating to trade, commerce and industry.

The Jackson Report, *Policies for Manufacturing Industry* (October 1975), pointed out that in Australian manufacturing, 50 per cent of all 'value added' is produced by 200 large, capital-intensive firms employing 44 per cent of the manufacturing labour force, and the other 50 per cent by 30 000 small and medium, labour-intensive firms employing 56 per cent. This division may lead to a Catch-22 situation where the management of smaller firms, in order to avoid extinction or take-over, feel that they can only survive by shedding labour and adopting computerized high technology. To stay alive, small firms may have to become even smaller as employers. But in an increasingly interdependent global economy, even the industrial giants may be at risk: BHP, for example, was forced to close its ship-building works at Whyalla in the face of impossible competition from the Republic of Korea. Multi-national corporations can transfer their operations to Singapore, Hong Kong, or the Philippines at short notice to find cheaper labour, even in manufactures where the size of the labour force has been of diminishing significance for decades.

Leaders in agriculture and mining press for lower tariffs because their products are in sufficient volume and quality to compete on the world market. If we wish Japan to buy more of our primary products, the Japanese – not surprisingly – feel that we should reduce or eliminate barriers against the free entry of their manufactured goods. It is axiomatic that the most protected manufacturing industries are the most labour-intensive: in fact 'labour-intensive' and 'protected' are synonyms. Similarly, 'low productivity' means 'high labour content relative to output', and 'productivity increase' is a euphemism for 'labour shedding'. In July 1973 the Whitlam government made an across-the-board tariff cut of 25 per cent, hoping that import competition would reduce inflationary pressures by forcing down the price of locally made goods. The boldness of this gesture caused shock waves not only throughout the business community but within trade unions and the Australian Labor Party (ALP), traditionally a high-tariff party despite the breadth and altruism of its foreign policies. It is hard to disagree with the view expressed by journalist Maximilian Walsh that

> the Whitlam decision put the tariff debate back years because it transformed a complex and exceedingly sensitive area of policy into an emotional political cause exploited with much manipulative skill and high personal gain by some very astute individuals and firms.

The textile industry reached its employment peak of 57 500 workers (1·22 per cent of the total labour force) in July 1965 and has declined both absolutely and relatively since then, falling to 37 200 (0·59 per cent) in 1978. The 1973 tariff cuts certainly accelerated employment losses, but the Industries Assistance Commission (IAC) attributes the decline to many causes – particularly changing patterns of demand. Employment in the clothing industry fell from a peak of 95 900 (1·8 per cent) in 1971 to 62 200 (0·98 per cent) in 1978, but this was only partly attributable to low-cost foreign imports. Some were cheap high-volume lines, but many were expensive high-quality goods. The IAC probably points to reduced expenditure in clothing as a proportion of the household budget. The significant drop in employment in the clothing industry over the period of 1973 cuts (1972: 93 200; 1975: 75 100) must be seen in the context of a long-term trend. Employment has fallen more dramatically in electronics manufacturing, from 30 490 (0·51 per cent) in 1972 to 20 882 (0·33 per cent) in 1978. Far more people work in the textiles, clothing and footwear industry – where export prospects might be charitably described as doubtful – than in the mining sector which is export-oriented. If Australia decides to opt for high technology and take on the international market (e.g. by large-scale metals fabrication), the price will be paid first by the migrant women who lose their jobs in protected industries.

It will be extraordinarily difficult for governments to pursue industry policies which are half adventurous and half protective, half no-tariff and half high-tariff. Whatever the case, the new high-technology export-oriented industries will be capital-intensive, providing a relatively small number of highly paid jobs.

Dr Fred Emery of Canberra has argued that the productive capacity of many Australian factories is already far in excess of the requirements of the local market and that, even if domestic demand rose sharply or export markets opened up, increased output could be achieved without further employment or acquiring new plant.[14] (This is certainly true of the Australian motor industry.) In 1891 Australia's manufactures employed one worker in five, in a relatively simple society with a limited range of readily available products. Ninety years later, the range of products is enormous and manufacturing still employs one worker in five. The development of new products has not in the long term raised the relative size of the sector – they have, however, prevented it from disappearing altogether. Dr Jonathan Gershuny argues in *After Industrial Society?* (Macmillan, 1978) that increasing dependence on manufactured goods may result in the loss of more jobs in services than in manufacturing, e.g. if more people buy their own lawn-mowers instead of paying the gardener, if they acquire handtools as part of the current enthusiasm for DIY activities, or if they abandon public transport and rely on their own cars. There are several likely areas for future expansion in manufacturing, such as solar-energy equipment, heat pumps, pleasure craft, or computerized toys. However, tariff reductions or any move towards freer international trade or large-scale rationalization could destroy employment in textiles, clothing, footwear, motor vehicles and accessories. It is difficult to see much opportunity for net expansion in this area.

Much of the variation in the secondary lines in the four- and five-sector graphs on pp. 60 and 61 is caused by fluctuations in the construction industry. In 1881, construction accounted for 7·09 per cent of Victoria's paid labour force; in 1891, the Australian figure was 7·86 per cent; and in 1901, 5·72 per cent. It rose to 8·75 per cent in 1976, and fell to 7 per cent by 1979. The building industry pioneered many labour-saving devices, and the construction of multi-storey buildings requires high technology – new methods have reduced the number of jobs dependent on physical strength alone. This has threatened the employment prospects of many male migrants, large numbers of whom were brought to Australia to work on construction projects which are now carried out by more sophisticated techniques. Much building activity occurs in the informal economy – where owners renovate old buildings as a hobby or an investment, or both. It is easy to think of many public projects – hospitals, rebuilt schools, welfare housing – which use intensive construction activity, but it seems doubtful if this would necessarily restore the industry to its 1976 levels of employment.

Tertiary (tangible economic services)

Economists have been inclined to disregard the employment significance of services, regarding them merely as links between the producer of goods and the consumer. However, the provision of tangible services – transport, storage, retailing, utilities, repair and maintenance, police and armed forces, and many personal services – has always been a very large employer in Australia. In Victoria the tertiary sector employed one paid worker in three as early as 1881, and this proportion has remained virtually unchanged to the present day. Tertiary employment (excluding construction) has generally employed more people than agriculture and manufacturing combined, reaching a high point in 1979 at 32·7 per cent of the paid labour force. However its growth seems likely to taper off in the 1980s, stabilize and probably fall. New technologies have reduced employment in retailing, storage, and fuel supply. Use of public transport in most states has fallen sharply, and may require a chronic fuel shortage to reverse the trend.

In the United States, Britain and France significant numbers are employed in unskilled, poorly paid and relatively low-status jobs in the tertiary sector – shoeshine boys, street sellers, attendants, janitors, *concierges* – but this type of employment has never been common in

Australia and is unlikely to be introduced now. Many craft skills may disappear. Television repair and maintenance workers are a vanishing species: who would bother to have a transistor radio repaired? Conventional watch maintenance was complicated and expensive – but quartz-digital watches have no moving parts to adjust or clean. Modular units in carpentry and plumbing make it cheaper to replace a defective fitting than have it repaired.

The most likely areas of expansion in tertiary employment are domestic tourism, leisure and sport. There are, however, four preconditions:

1 Maintenance of equitably spread income levels.
2 Greater blocs of usable time – such as a shorter working year.
3 Government support for the provision of leisure facilities and ancillary services: promotion, information, transport service, recreation areas.
4 Australia would need to entice more tourists from abroad or persuade more of its citizens into domestic travel.

In Europe, Africa and Asia tourism has grown enormously in the past twenty years. Australia still lacks competitive appeal – the basic fares from Europe to Australia are very high, and the great distances between major attractions such as the Barrier Reef, Kakadu National Park and Sydney are further complicated by very expensive domestic air fares.

Sport has enormous potential as an employment absorber. In his *Green Paper on Sport and Recreation* (1980), Barry Cohen pointed out that Australia has an annual health bill of about $8000 million yet government spending on sport, physical-fitness campaigns and promotion of alternative lifestyles (less passive, less drug-dependent) amounted to less than $4 million for 1979–80. He also drew attention to the success of the *Spartakiad* programme in the German Democratic Republic, which involves a high level of direct community involvement in sport – not only in time absorption by amateurs, but also in the employment of coaches and administrators, and people building stadia, making sporting equipment, and maintaining sports grounds.

Quaternary (information processing)

Politicians, economists, public servants and media people are all in this sector – but it is slow to be recognized as a separate or significant entity. To many people, work is essentially what other people do. In 1881, 5·5 per cent of Victoria's paid labour force worked in information. By 1976, it had become the largest Australian sector (30·9 per cent), only to fall slightly behind the tertiary sector in the 1979 figures, the first indication of the 'bite' of new technology.

The printing industry is marked by falling numbers and vastly increased productivity: the membership of the Printing and Kindred Industries Union (PKIU) fell from 59 159 in 1974 to 51 675 in 1978, despite a large increase in both the volume of printed material and the size of Australia's labour force generally. It has been estimated that to produce a single page of text in a large broadsheet newspaper might take a hand compositor twenty-two hours, a machine compositor 5·5 hours, a TTS (teletypesetting) casting unit 1·3 hours – and an electronically controlled photocomposing machine fifteen seconds. The computer industry itself is often cited as a major growth area in employment. Its growth rate from a zero base is impressive, but it seems likely that for every job created by the computer industry about two will be eliminated.

In the global economy, the United States is still head office. When the United States carries out basic research and development in vast industries such as aerospace, telecommunications, computers, or defence hardware, it does so for the entire world. Development of the new technology appears to create jobs in the US and displace them elsewhere: as a result, employment growth in teaching, banking, insurance, telecommunications and office work in Australia is increasingly doubtful. In any case, Australia lacks the necessary educational base to follow the United States into the 'knowledge society': it is no coincidence that the only nation with almost 50 per cent of its population in the information sector is also the only nation with over 50 per cent of young adults undergoing tertiary education (see also Chapter 7).

Quinary (domestic and quasi-domestic servicing and/or making)

Since Palaeolithic times, unpaid domestic work has always been the first or second employment sector, a fact ignored by economists and statisticians. It is now slowly being recognized that these services have enormous direct and indirect economic significance: they are central to the functioning of the community, direct the movement of production and consumption, do a great deal of direct production and are essential for population growth.

Estimates taken from the 1881 Victorian census for occupations notionally included in my quinary sector – including as it does paid workers in accommodation, the service of food and drink, cleaning and some handcrafts – suggest a figure of 29·11 per cent, surprisingly low. This confirms that very many married women worked in industry: even in 1881 there were twice as many women in the textile industry as men. The graph on p. 61 illustrates a steady rise in the size of the quinary sector between the censuses of 1911 and 1947, as more women left the paid labour force and stayed at home, especially during the 1930s depression. The category of 'women who kept house without wages', as they were usually defined, was at its peak in the 1947 census. The graphs are accurate in indicating long-term general trends, and their lines link figures taken from census returns, but as there was no census between 1933 and 1947 they give the misleading impression that there were only steady changes in sectoral rises in that period. During World War II, however, women joined the armed forces, worked in heavy industry and on farms. When the war ended, some wished to return to domesticity and 1947 was the year of our highest birth rate. (In absolute figures there was only a small decrease in the number of working women, due to rising population.)

Unpaid domestic work is ignored in the calculation of the national accounts. The economist A.C. Pigou illustrated this with his famous story of the widowed clergyman whose housekeeper's wages were included in the national accounts until he married her, after which she ceased to have any statistical significance. It is very difficult to calculate the work value of unpaid domestic labour, but if the services provided in most households had to be paid for on a professional basis there are few breadwinners who could meet the salary claim of their spouse.

There is a growing amount of paid employment in the quinary sector, e.g. it is now widely accepted that there should be paid home help for the aged, sick and disabled. This indicates that society acknowledges the need to support home life, but these services do not do so merely in the traditional sense of 'keeping women in the home', they also provide a widening of opportunities: with child-care services and various types of domestic assistance, more women have greater freedom to enter the paid labour force, especially in part-time work (married women comprise 64 per cent of part-time workers in Australia). The quinary sector fell consistently from 1947 to 1978 because an increasing number of married women were entering the paid labour force, as illustrated by the diagram on p. 61. However, the upward turn in the 1979 figures in this graph indicates that the post-war trend may be reversing itself.

THE CHANGING STATUS OF DOMESTIC WORK

Hugh Stretton has pointed out that:

> In affluent societies (as in most others) much more than half of all waking time is spent at home or near it. More than a third of capital is invested there. More

Male / female participation in the Australian labour force, 1947–76

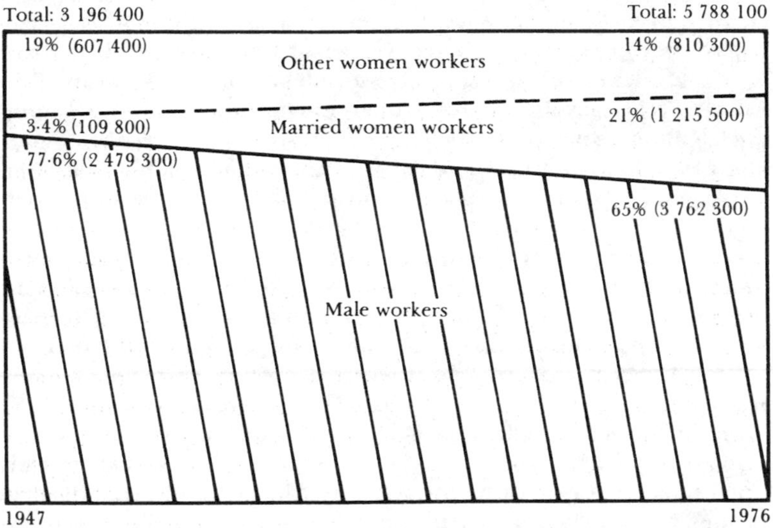

than a third of work is done there. Depending on what you choose to count as goods, some high proportion of all goods are produced there, and even more are enjoyed there. More than three quarters of all subsistence, social life, leisure and recreation happen there. Above all, people are produced there, and endowed with the value of capabilities which will determine most of the quality of their social life and government away from home. So the resources of home and neighbourhood have a commanding importance . . . It is in the activities of home, neighbourhood and voluntary association that there is least money exchange, least division of labour, least bureaucracy, least distinction between production and consumption, least occasion for oppressive or exploitative or competitive uses of ownership, and most of the best opportunities for cooperative, generous, self-expressive, *unalienated* work and life. Anyone interested in building a more cooperative, affectionate or equal society should therefore look first, and centrally, to the resources of home and neighbourhood.

But that whole domestic economy – a third or more of every developed economic system and two thirds or more of every social system – tends to be systematically undervalued and consequently starved of resources by the orthodox economic thinking of both Left and Right . . . It should be socialist policy to shift more resources into the domestic sector, and as far as possible to equalise household capital.[15]

Australia has about 1·5 million families in which both husband and wife work – and the proportion of two-income families has increased steadily since the 1960s. Increasing labour-force participation by married women has coincided with growing numbers of unemployed youth, and many have wrongly concluded that the expansion of one has caused the contraction of the other. However, the Australian labour force is still sexually segmented to a high degree, and the area where large numbers of married women are employed – office work, teaching, sales – are not responsible for non-employment of unskilled workers, especially males (nor would removing married women from the labour force provide work for those presently unemployed).

US Department of Labor statistics confirm that in the 1970s great employment growth took place in my 'quinary sector'. With increasing proportions of women entering paid work outside the home, there has been a corresponding transfer of some formerly domestic functions (e.g. family eating) to the market economy. This trend may not last, and the effect of long-term economic depression and increasing unemployment will be felt first and most severely in the quinary sector. Apart from economic insecurity, quinary employment provides jobs, not careers, largely 'dead-end' and unrewarding in esteem, promotion or pay. While there is little evidence that married women displace the unemployed young, it is possible that significant work opportunities can be created for highly paid domestic services performed on a contract basis. For many working wives, routine domestic work such as cleaning, shopping, preparing meals, washing clothes, taking children to school and picking them up, must be carried out in addition to full- or part-time work. The British sociologist Michael Young has pointed out that in Britain many young couples find full- or part-time contract domestic work remunerative and professionally satisfying. It would, for example, be comparatively easy for one couple to service five two-income families in a particular locality on a regular basis: to rationalize the supply of meals, take children to and from school, and use high-quality equipment to mow lawns and sweep carpets. The couple could earn good money and would have considerable variety and flexibility in arranging their own work schedules.

Michael Young and Peter Willmott, in *The Symmetrical Family*, have noted changes in the traditional model of 'husband breadwinner' and 'wife homemaker' and see the evolution of an arrangement where both partners do some paid work and share the domestic work. This could lead to a considerable reduction in the working hours of both parties, leading to a greater sharing of home activities. With increased leisure time, and more part-time work, this tendency will be much more common. The best result of these changes could be that individuals find themselves mostly occupied doing things they *like* rather than those they are condemned to by financial need or traditional sex roles.

Changing the status of domestic work may provide wider real options for many people – mostly but not exclusively women – who are forced by economic circumstances to seek jobs in manufacturing or services but who might prefer to work at home if the family could afford it. For example, people who undertake the care of the aged in their homes – a problem which will grow in an increasingly geriatric community – ought to be paid not less than the government subsidy to old peoples' homes for each inmate. This would involve extending family allowances in recognition that internally provided (domestic) services may be substituted – where appropriate – for externally provided (professional) services. In the UK, certain 'contra allowances' are made in respect of the married woman worker – presumably to help pay for home help, child-minding, etc. In France the household income is added together and then divided by all household members, which creates a benefit for large families. In the United States, although income tax is a Federal matter, its mode of collection varies between states, and many allow for income splitting. In the states which were influenced by French or Spanish law it was assumed that in a single-income family wages were shared equally by husband and wife.

The economic and social value of domestic work must be recognized for its own sake, and not just as a device to keep married women out of the paid labour force. Similarly, the importance of do-it-yourself work – including home maintenance – should be acknowledged and its value added to the national accounts. Study and much non-economic activity such as home-based craftwork should also be recognized as economically and socially useful: a student may prove to be a valuable physician, and a recluse may produce a brilliant novel. (James Thurber used to say, 'I could never persuade anyone that I was working when I was gazing out the window.') Further, with the increasing use of technologies such as Ceefax or Teletext, home computer terminals and picturephones, a great deal of work which is now performed in offices can just as easily be performed at home – at far lower cost, and with enormous saving of time and transport costs. Ordering goods directly from the warehouse can be done by telephone or computer: so can banking, paying bills, or gambling. In a post-service society, domestic work may once more become the norm. Work in the home has been grossly undervalued, and raising its status might be an important first step to a more humane and co-operative society, which will replace the present competitive model where too much emphasis is put on economic activity external to the home and on the compulsory work ethic. (The work ethic, and the need to recognize the value of time use outside paid employment, are examined in Chapter 9.)

In Sweden, more than 20 per cent of urban families have a second home, usually in the countryside, and a considerable distance from the first home.[16] The second home represents a major capital investment, and large slabs of time are used in maintenance and in travelling between the two homes. Frequently the second home provides 'off-peak' occupation for aged relatives – it is an option which may become increasingly popular as a means of labour and time absorption. In Australia the 1976 census indicated that 2·2 per cent of dwellings were second or holiday homes: however, this figure is almost certainly understated.

CASE STUDIES: KOOYONG, LALOR AND MALLEE

To understand how structural change in employment is occurring, it is useful to look at disaggregated statistics for particular regions. Aggregated statistics often mislead – employment figures for apple growing and sugarcane farming have enormous significance in Tasmania and Queensland, but adding them together and dividing by six states to obtain a national figure would be pointless. Similarly, a decline in manufacturing employment is partly concealed when figures for areas dominated by manufacturing are lumped together with areas where manufacturing has never been a significant employer. Variations in educational participation or youth employment rates can only be grasped by examining figures from particular regions.

Three of Victoria's thirty-three seats in the House of Representatives – Kooyong (an affluent, highly educated, suburban electorate), Lalor (a sprawling outer-urban industrial seat with a heavy concentration of migrants), and Mallee (a large rural seat, dominated by agricultural employment) – provide useful illustrations of diversity in employment and the impact of structural change. *Kooyong* is a safe Liberal seat in Melbourne's inner eastern suburbs (Auburn, Balwyn, Camberwell, Hawthorn and Kew), a compact (42 square kilometres), long-established residential area. It has good public transport, easy access to the city centre, very high land prices, good community services, and 13·61 per cent of its population was born in non-English-speaking countries. *Lalor* is a safe ALP seat comprising Melbourne's outer western suburbs (Sunshine North and West, St Albans, Deer Park, Altona and Laverton), the town of Werribee, and four rural hamlets in an area of 817 square kilometres extending towards Geelong. It is a mixture of heavy industrial and residential areas which expanded rapidly after World War II, with inadequate public transport and poor internal communication, comparatively cheap

Changes in the secondary sector (manufacturing and construction) by percentage

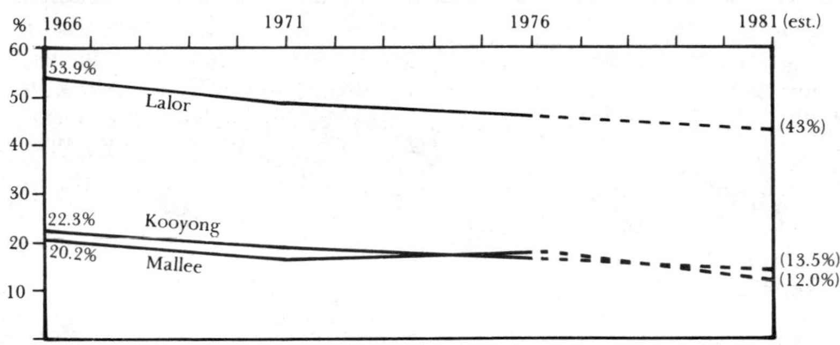

Changes in the quaternary sector (information) by percentage

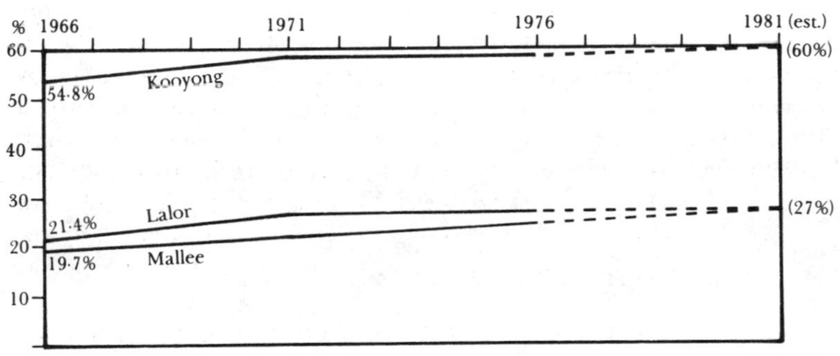

land, poor provision of services, and with 29·27 per cent of the population born in non-English-speaking countries (mainly Yugoslavia, Malta, Italy and Greece). *Mallee* is a safe National Country Party (NCP) seat in the north-west of Victoria, bordering on New South Wales and South Australia. It is a rural area based on wheat, wool, fruit, vineyards, cattle and dairying, including the towns of Mildura, Swan Hill, Horsham, Warracknabeal, Hopetoun and Nhill, widely dispersed (57 720 square kilometres), with cheap land, and 4·05 per cent of the population born in non-English-speaking countries.

Kooyong is moving from a post-industrial society to a post-service society, and is a model of Australian occupational patterns for the 1980s. Lalor is still a model of Australian industrial society of the 1950s, largely unaffected by the middle-class educational revolution of the period 1965–75 and the consequent stimulation of service employment. Mallee is a model of pre-industrial society moving

Profiles of three electorates: Kooyong, Lalor and Mallee

People	*Kooyong*	*Lalor*	*Mallee*
Electors enrolled (1980)	71 792	80 522	67 763
Population	103 188	122 055	103 410
Children aged 0–16 years	23 157 (22·44%)	44 765 (36·67%)	34 633 (33·49%)
Not in labour force (aged 15+ years)	33·7%	21·6%	26·0%
People aged over 16 years born in non-English-speaking countries	17·54%	46·23%*	3·78%
People who took holidays in census year	57·7%	35·6%	53·0%
People in government employment	23·6%	23·9%	19·8%
Aged 65 years +	4·3%	0·6%	10·8%
Employment (%)			
Agriculture and mining	0·8	2·2	35·2
Production process workers	17·1	46·2	17·3
Armed forces	0·3	4·5†	0·1
Transport and communications	3·1	5·6	4·7
Sales	8·2	5·8	7·7
Service, sport and recreation	6·9	6·4	6·9
Administration and clerical	32·7	18·2	13·8
Professional, technical, teachers	25·4	6·3	8·8
Others	5·5	4·8	5·5
Income, 1976 (%)			
More than $18 000	24·8	13·1	5·3
$9000–$18 000	31·9	43·5	24·5
Less than $9000	34·5	28·9	61·7
Not stated	8·8	14·5	8·5

* This understates the number of people in Lalor whose primary language is not English: the Australian-born children of Greek parents, for example, will often follow the family language.

† Lalor includes the Royal Australian Air Force base at Laverton-Point Cook.

towards industrialism, and also remote from changes in education. The 1976 census figures (corrected to 1977 boundaries) give the profiles set out in the table above (for education figures, see Chapter 7). The political allegiance of each electorate, as shown in the 1980 general election results, accurately reflects its employment base. (The 'two party preferred' vote is the estimated result if all votes had been notionally distributed among two parties only.)

Sectoral change in specific seats: 1966, 1971 and 1976 censuses

	1966		1971		1976	
		per cent		per cent		per cent
Four-sector analysis						
Kooyong:						
Primary	403	1·0	342	0·8	361	0·8
Secondary	9376	22·3	7937	19·0	8126	17·1
Tertiary	9188	21·9	9204	22·0	10 993	23·2
Quaternary	23 031	54·8	24 353	58·2	27 911	58·9
	41 998	100	41 836	100	47 391	100
Lalor:						
Primary	1346	2·8	1159	2·1	1169	2·2
Secondary	26 246	53·9	26 566	48·0	24 614	46·2
Tertiary	10 644	21·9	13 117	23·7	13 426	25·2
Quaternary	10 446	21·4	14 505	26·2	14 068	26·4
	48 682	100	55 347	100	53 277	100
Mallee:						
Primary	13 612	40·3	11 611	36·9	15 513	35·2
Secondary	6818	20·2	5637	17·9	7644	17·3
Tertiary	6707	19·8	7133	22·6	9789	22·2
Quaternary	6665	19·7	7113	22·6	11 167	25·3
	33 802	100	31 494	100	44 156	100
Five-sector analysis						
Kooyong:						
Primary	403	0·8	342	0·7	361	0·6
Secondary	9376	18·0	7937	15·27	8126	13·8
Tertiary	8223	15·8	8237	15·9	9947	17·0
Quaternary	23 031	44·1	24 353	46·8	27 911	47·4
Quinary	11 119	21·3	11 082	21·3	12 506	21·2
	52 152	100	51 951	100	58 851	100
Lalor:						
Primary	1346	2·1	1159	1·6	1169	1·7
Secondary	26 246	41·7	26 566	37·1	24 614	35·7
Tertiary	10 016	15·9	12 343	17·2	12 674	18·4

	1966		1971		1976	
		per cent		per cent		per cent
Quaternary	10 446	16·6	14 505	20·2	14 068	20·4
Quinary	14 963	23·7	17 073	23·9	16 443	23·8
	63 017	100	71 646	100	68 969	100
Mallee:						
Primary	13 612	32·0	11 611	29·2	15 513	27·9
Secondary	6818	16·0	5637	14·2	7644	13·7
Tertiary	5387	12·6	6358	16·0	8830	15·8
Quaternary	6665	15·6	7113	17·9	11 167	20·1
Quinary	10 121	23·8	8976	22·7	12 502	22·5
	42 603	100	39 695	100	55 656	100

Actual vote (%)

Kooyong		*Lalor*		*Mallee*	
Liberal	57·6	ALP	68·4	NCP	47·2
ALP	33·9	Liberal	25·6	ALP	23·9
Others	8·5	Others	6·0	Liberal	21·4
				Others	7·5

Two party preferred (%)

Liberal	61·5	ALP	71·5	NCP	70·8
ALP	38·5	Liberal	28·5	ALP	29·2

The table on pp. 75–6 applies four- and five-sector analyses of the labour force to the census returns for 1966, 1971 and 1976. This and the accompanying graphs illustrate the enormous contrast in employment patterns in the three seats and the degree of structural change over a decade. (Electoral redistributions in 1968 and 1977 changed the boundaries of these seats substantially, but not their economic and political character. The 1966 and 1976 census figures have been corrected for the new boundaries.) In Victoria, 20·5 per cent of the labour force is employed in manufacturing: Lalor's industrial employment is more than twice the state-wide percentage. Lalor is marked by employment which is not particularly responsive or sensitive to demand, e.g. it is the centre of Australia's petrochemical industry where output increases dramatically while employment falls steadily. There is little prospect for increasing the number of industrial jobs in Lalor – it goes against a historic trend, the capital required for each new job is very high, and the cost in pollution would be excessive. On the other hand, expanding services such as retailing, construction, education, recreation, and the supply of food and drink would create many new jobs. Lalor has more people employed, both absolutely and relatively, than Kooyong – but it also has far more unemployed, and is incapable of absorbing people in non-paid occupation (such as study) as is the case in Kooyong.

Lalor illustrates the problems caused by failure to adjust to structural change and the inevitable limitations of a restricted educational base. If it is taken for granted that boys leaving school will work in factories and girls in supermarkets – both being cut back by new technology – then youth unemployment must increase. Lalor cannot easily make the transition to a service or post-service society because it lacks the necessary education base. Families in Lalor who want to climb up the socio-economic pyramid may find it easier to do so by moving out of the area altogether. Mallee is also crippled by its narrow educational base. It seems likely that young people seeking further education and social mobility will move away to the major cities – that the range of work opportunities will contract. Kooyong is dominated by information employment which is, at least for the time being, labour-intensive and demand-sensitive. However, it seems likely that the information sector has already peaked as an employer: Kooyong faces a period of adjustment to new forms of employment as well.

WHITE COLLARS AND BLUE COLLARS

The aggregated figures for white- and blue-collar employment in Australia, as set out in the simplified diagrams on p. 78, demonstrate the trends over a thirty-year period. The trends may continue, but we cannot be sure: it seems likely that white-collar clerical work will fall rapidly, and process work in manufacturing seems virtually certain to fall even if tariff protection props it up. The schemas also illustrate how it is possible to make two seemingly contradictory assertions based on the same data:

1 Australia has more blue-collar workers (in manufacturing, repair and maintenance, heavy transport, utilities, mining, farming) than ever before.

2 Australia has a smaller proportion of blue-collar workers now than at any time in its history.

Both are correct. The first proposition depends on absolute figures, the second on proportions.

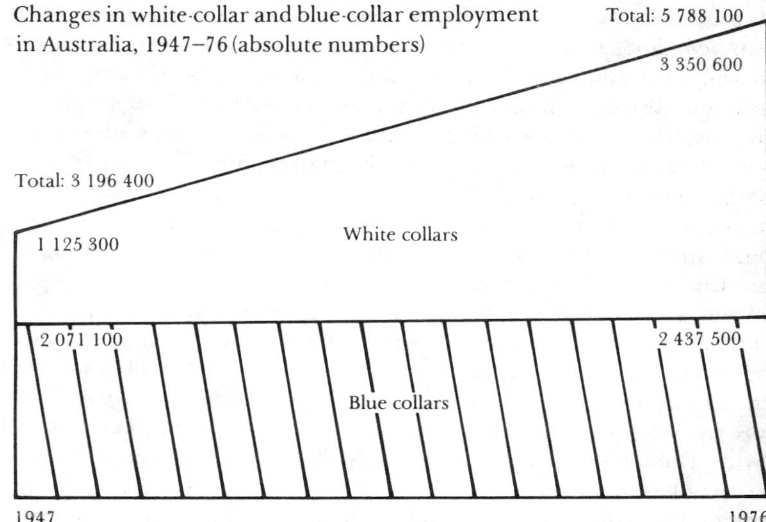

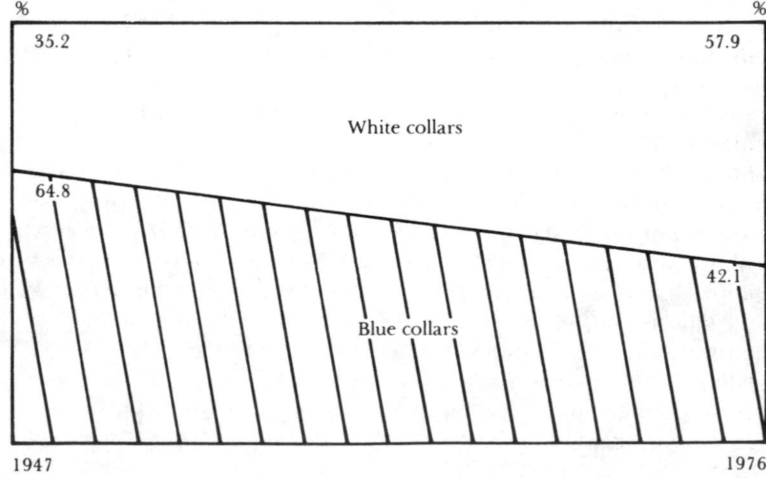

Conclusion

Manufacturing will continue to decline as a proportion of the labour force. The rate at which this occurs will be affected by both the maintenance (or withdrawal) of tariff protection for labour-intensive industries and the speed with which Australian or foreign capital relocates overseas, but the trend will not change. New capital-intensive industries such as aluminium processing and metals fabrication will be established in Australia, but they will be insignificant employers. The best hope for employment growth in the secondary sector would be in the revival of craft-based employment where work is in itself an output of production – skilled trades such as carpentry, tailoring, jewellery, medical prosthetics, pottery, cutlery – or where it is not subject to direct competition from technology or imports, is not capital-intensive, and is based on the satisfaction of individual needs. Such work will inevitably be labour/time-absorbing, low in productivity, and consequently high in price. We must dispel the illusion that vast numbers of new jobs can be found in high-volume production.

White-collar work in the information sector is also at risk because of the low cost and extreme portability of so much computer-based activity. The quinary sector seems to have the greatest potential for job creation – and the first step is to recognize the size and significance of the unpaid work already there. But this sector has enormous scope for fulfilling human needs on a continuing basis – for example, restaurants, entertainment and sex-related work. In the United States, between 1973 and 1980, three industries – 'eating and drinking places', 'health services' and 'business services' – increased employment three times faster than total private employment and sixteen times faster than employment in the goods-producing sector: 'The *increase* in employment in eating and drinking places since 1973 is greater than total employment in the automobile and steel industries combined.'[17] In 1979, Americans paid $87 000 million to eat in the nation's 320 000 restaurants – the equivalent of almost $20 per head for every person on earth, and more than the combined worldwide earnings of Exxon and IBM. In this area, the capital cost of job creation is very low (often less than $10 000 per job) compared to heavy industry such as aluminium smelting where the capital cost per job in Australia is estimated to be $933 000 (1981).

Administration, education and welfare still have considerable scope for employment generation – but outside the market economy. Some labour-intensive industries have become *de facto* welfare institutions in which income is redistributed to help maintain employment levels for their own sake – and the end product is not so much shoes and singlets as welfare.

4 Two Types of Employment and Time Use

Q. How many people work in the Vatican?
A. About half.

Answer attributed to Pope John XXIII

There are two basic and fundamentally contradictory forms of employment and time use in modern society:

1 labour/time-saving;
2 labour/time-absorbing.[1]

In Volume I of *Capital*, Karl Marx distinguishes briefly between 'dense' and 'porous' employment, and refers to

> the compulsory shortening of the hours of labour. This gives an immense impetus to the development of productivity and the more economic use of the conditions of production ... The denser hour of the 10-hour working day contains more labour, i.e. expended working power, than the more porous hour of the 12-hour working day.[2]

This distinction is implicit in his *Grundrisse*, Notebook VII.

Labour/time-saving work (usually called 'capital-intensive') is characteristic of the division of labour and technological efficiency in agriculture, mining, manufacturing construction and some services (e.g. retailing, storage, or bulk transport). It is directed towards the maximum output of some tangible 'good' (or maximizing access to it) for a minimum investment of labour and time. Its success is measured objectively – by market profitability, 'cost-efficiency' and productivity (a relative test). If two factories making singlets have an identical labour force but the productivity of one is greater (e.g. 50 per cent more garments each day) this will be reflected in its lower prices: it has higher cost-efficiency and the capacity to meet greater demand. Success is ultimately measurable by output relative to the total input – i.e. 'productivity'. With new employment patterns, the development of a global economy, and adoption of new technology, a decreasing proportion of Australians need to be employed in labour/time-saving work.

Labour/time-absorbing work (usually called 'labour-intensive') is

Two types of work

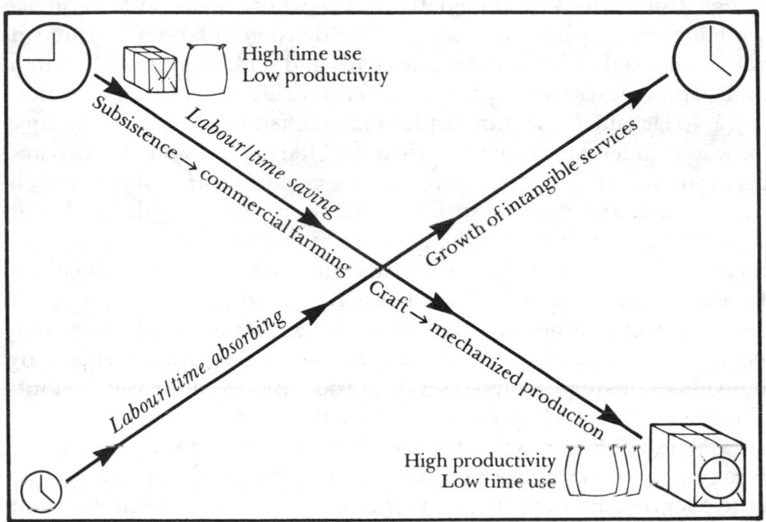

Three economic and time-use eras

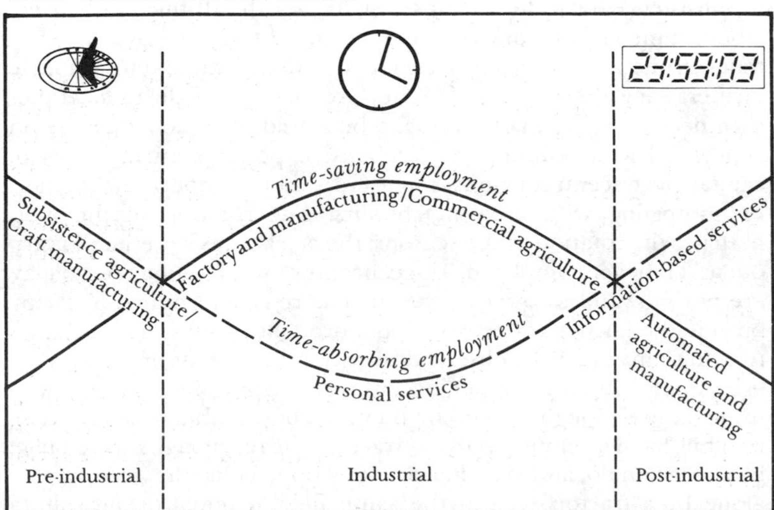

characteristic of most service employment in sophisticated economies. Historically it included those for whom work and existence were inextricably linked, unaffected by 'division of labour': farmers in subsistence agriculture, domestic work and child-rearing by women, and people devoted to a vocation-centred life where market forces were irrelevant (religious, traditional craftsmen, poets, musicians). Now it includes education, health and welfare, provision of meals and accommodation, information services, administration, research, tourism and entertainment. Its products are increasingly costly and their 'value' notoriously difficult to measure, except subjectively. An increasing number of Australians are employed in such occupations. Productivity cannot be high, as many of the services are high in cost and low in output because the ratio between servers and served may be very low (e.g. in dentistry or teaching, where the receivers are individuals or small groups). Service productivity is not easily quantifiable as yet, but it is clearly not matching the productivity growth in the areas of tangible goods production. (In social-welfare terms, however, it may well be that services productivity *is* growing.) The test of 'cost-efficiency is only peripheral to labour/time-absorbing employment. When Henry Moore was a young sculptor his work had no commercial value whatever, but now those stones which the connoisseurs rejected are enormously valuable: an accountant in 1930 would have rated Moore's 'productivity' as low, while fifty years later we would say it had been very high. But the tests of commercial success are applied at the time of operation, not after the event.

Subsistence agriculture helps to illustrate the distinction between labour/time-saving work and labour/time-absorbing work. In subsistence economies, agriculture is not regarded as an 'industry' for the farmers – it is their way of life. They produce essentially to meet their own needs, with a small surplus to be stored or traded; they are not concerned with economic profitability, export potential, return on capital, or concentrating on a single crop; they are not racing the clock or competing with their neighbours. The rotation of the crops matches the rotation of the seasons; the work takes a lifetime to carry out and is never completed. The concepts of wages, hours or holidays are not relevant. Relative success or failure is determined by factors outside the farmer's control – flood, drought or other natural disasters. In China in 1978, I observed large numbers of commune workers – between 150 and 200 per hectare – ploughing fields. Some were using simple ploughs pulled by human labour, others used mechanized but hand-operated 'walking tractors', heavy ploughs were pulled by water buffalo, and even heavier steel-tipped ploughs were dragged along by a tractor – all in the same field system. An agricultural economist assessing the 'cost-effectiveness' of ploughing by tractor power, water-buffalo power and human muscle power might have said: 'Properly mechanized, you can dispense with 80 per cent of the commune's workforce.' But the commune's leaders would have replied: 'What will they do then?' The economic systems of the USSR and China are saved from high levels of unemployment by the relative inefficiency of their agriculture, which absorbs 22 per cent and 65 per cent of their respective labour forces. (The same applies to India, where the figure is over 60 per cent.) If, like the United States, they fed their people with an agricultural sector employing only 3·5 per cent of the labour force they would face massive social problems.

Mexico is currently passing through a very rapid agricultural revolution. Due to the enormous increase in population (3·9 per cent per annum), the demand for food can no longer be met by subsistence agriculture. Mexican farm labourers are said to have worked for an average of only 135 days every year: since the traditional work ethic had less significance in Mexico than the heat, the fiesta, the siesta and *mañana*, Mexican rural life had a slow and deliberate rhythm. But the consolidation of land holdings, mechanization, aerial spraying and other forms of modern farming which are converting Mexican agriculture into a capital-intensive industry make the *campesino's* labour irrelevant – and displaced farm workers are streaming into Mexico City at an average rate of 3000 every day. A rigorous application of cost-effectiveness to subsistence agriculture would lead to massive unemployment in Asia and South America, and increase the already serious movement of population from the countryside to the cities.

My (belated) insight was to recognize that labour/time-absorbing employment was the norm in human experience until the eighteenth century: the agricultural revolution in Europe and North America converted agriculture from a life-absorbing experience to an increasingly capitalized, mechanized and specialized industry; the Industrial Revolution had the same effect on craftwork. The worldwide growth of service employment after World War II reversed this trend, and greater inputs of time and labour resulted in smaller outputs. However, the development of computerized machines possessing something analogous to intellectual skills, and tireless and faster than their human rivals, raises a challenge to service employment that it has never faced before. The technical capacity exists to produce low-cost high-quality services which until now required heavy investment in human time and skill. We appear to be in the first stages of a 'post-service' revolution, in which machines will do to human brain power what the First and Second Industrial Revolutions did to human muscle power.

Theories of work and time use

The complex relationship between labour/time-saving work and labour/time-absorbing work can be illustrated by reference to the writings of Adam Smith, C. Northcote Parkinson, and those who adapted their theses. (The philosophies of work propounded by Smith, Marx and others, and the historical impact on work of the division of labour, are discussed further in Chapter 9.)

SMITH, BABBAGE AND TAYLOR

Adam Smith was the St Paul of labour/time-saving employment, and although he did not use the term 'economies of scale' (which includes the concept that fewer specialized workers could effect an enormous increase in output) it is implicit in his famous advocacy of the division of labour which opens Book I of *Wealth of Nations*.[3] He illustrated the division of labour by the example of pin-making: 'an untrained workman . . . with the utmost industry . . . could scarcely make one pin in a day and certainly not twenty' – but divide the work into about eighteen different operations and give it to ten workers, each performing two or three specialized functions, and the total output might increase to 48 000 pins in a day, an average of 4800 per man. 'The advantage gained by saving the time commonly lost in passing from one sort of work to another is much greater than might at first be imagined.'

> Everybody must be sensible how much labour is facilitated and abridged by the application of proper machinery. It is unnecessary to give any example. I shall only observe, therefore, that the invention of all those machines by which labour is facilitated and abridged seems to have been originally owing to the division of labour.[4]

In Book III Smith distinguished between 'productive labour' where work 'adds to the value of the subject on which it is bestowed', and 'unproductive labour' which 'has no such effect':

> The sovereign, for example, with all the officers both of justice and war who serve under him, the whole army and navy, are unproductive labourers. They are servants of the public, and are maintained by a part of the annual produce of the industry of other people. Their service, how honourable, how useful, or how necessary soever, produces nothing for which an equal quantity of service can afterwards be procured. The protection, security and defence of the commonwealth, the effect of their labour this year, will not purchase its protection, security and defence for the year to come. In the same class must be ranked some of the gravest and most important and some of the most frivolous professions: churchmen, lawyers, physicans, men of letters of all kinds; players, buffoons, musicians, opera-singers, opera-dancers, etc. The labour of the meanest of these has a certain value, regulated by the very same

principles which regulate that of any other sort of labour; and that of the noblest and most useful produces nothing which could afterwards purchase or procure an equal quantity of labour. Like the declamation of the actor, the harangue of the orator, or the tune of the musician, the work of them all perishes in the very instant of its production . . . Unproductive labourers, and those who do not labour at all, are all maintained by revenue.

Eli Whitney (1765–1825), the forerunner of modern mass-production techniques, devised a form of 'horizontal' division of labour which required only a few skilled workers, the rest being reduced to the status of 'hands' whose task was to feed material into machines where jigs guided the tools to make standardized, interchangeable products.

Charles Babbage, English mathematician and computer pioneer, in *On the Economy of Machinery and Manufactures* (1832) adopted Smith's distinction between 'productive' and 'non-productive' employment and extended the concept of 'division of labour' to mental work as well. He gave as example the work of French mathematicians who were directed by Napoleon to work out eight-figure logarithms of all numbers between one and 200 000 – and produced seventeen huge volumes in rapid time by dividing the work according to its complexity so that the least skilled mathematicians were performing the simplest tasks. He also recognized the importance of a 'horizontal' division of labour:

. . . the master manufacturer, by dividing the work to be executed into different processes, each requiring different degrees of skill or of force, can purchase exactly that precise quantity of both which is necessary for each process; whereas, if the whole work were executed by one workman, that person must possess sufficient skill to perform the most difficult, and sufficient strength to execute the most laborious of the operations into which the art is divided.[5]

Babbage uses the same example of divisions of labour as Smith – pin-making. His meticulous calculations show that a team of ten – four men, four women and two children – could produce 5546 pins in 7·6892 hours (about 8600 pins per day, less than one-fifth of Smith's estimate), where four skilled and six unskilled people worked together, where wage rates ranged from 3s 3d per day for the men and 1s for the woman, to between 4½d and 6d for the children. The American Marxist author Harry Braverman wrote of Babbage's principle:

Translated into market terms, this means that the labour power capable of performing the process may be purchased more cheaply as dissociated elements than as a capacity integrated in a single worker. Applied first to the handicrafts and then to the mechanical crafts, Babbage's principle eventually becomes the underlying force governing all forms of work in capitalist society, no matter in what setting or at what hierarchical level . . . This might even be called the general law of the capitalist division of labour.[6]

Babbage also wrote: 'One great advantage which we may gain from machinery is the check which it affords against the inattention, the idleness, or the dishonesty of human agents.' In other words, the economy, precision, regularity and tirelessness of the machine becomes the standard the human workers must also adopt.

The concept of 'scientific management', a logical extension of the ideas of Smith and Babbage on the division of labour and the concept of 'labour value', was developed independently in France by Henri Fayol (1841–1925) and in the USA by Frederick Winslow Taylor (1856–1915). Fayol insisted that 'to manage is to forecast and plan, to organise, to command, to coordinate and to control' and thought that running a business was like commanding an army: success depended on leadership from on high. Taylor took a different starting point. He was concerned to ensure that operating methods of the process workers in factories or offices were efficient, so that the greatest output could be achieved in minimum time and workers had no opportunity to loaf on the job. Taylor worked his way from machine-shop labourer to chief engineer of the Midvale Steel Company in Philadelphia, and then became a consulting engineer in management. He retired at forty-five to devote himself to promoting his ideas, and published *The Principles of Scientific Management* in 1911. 'Taylorism' is identified with time and motion studies, piece-work, the elevation of foremen to quasi-managerial status, incentive payments, and tools used in the most efficient manner. Braverman wrote of Taylorism: 'It investigates not labour in general, but the adaption of labour to the needs of capital. It enters the work place not as the representative of science, but as the representative of management masquerading in the trappings of science.'[7]

Taylor did not invent all the elements of scientific management, but he was able to synthesize and relate a series of minor changes and express them as 'scientific laws'. His ideas were adopted by the Bethlehem Steel Company, the Ford Motor Company and many major corporations. Taylor wrote that managers should

> assume . . . the burden of gathering together all of the traditional knowledge which in the past has been possessed by the workmen and then of classifying, tabulating and reducing this knowledge to rules, laws and formulae . . . All possible brainwork should be removed from the shop and centred in the planning or laying out department . . . The science of doing work of any kind cannot be developed by the workman. Why? Because he has neither the time nor the money to do it.

Taylorism represented an intensification of management control over workers, greater discipline, time domination, emphasis on productivity, and de-skilling process work – it was another blow to both the tradition of craftwork and the use of independent judgment in formerly skilled employment. However, the work of trade unions since Roosevelt's New Deal era has succeeded in limiting the significance of Taylorism in most US industries.

Taylorism was not confined to the capitalist world. V. I. Lenin specifically endorsed the adaptation of its principles to meet Soviet needs in three papers – 'The Immediate Task of the Soviet Government' (April 1918), 'Six Theses on the Immediate Tasks of the Soviet Government' (May 1918) and '"Left-wing" Childishness and the Petty Bourgeois Mentality' (May 1918) – and in one speech 'The Tasks of the Youth Leagues' (October 1920).

> We must raise the question of piece-work and apply and test it in practice; we must raise the question of applying much of what is scientific and progressive in the Taylor system; we must make wages correspond to the total number of goods turned out, or to the amount of work done by the railways, the water transport system, etc., etc . . . The task that the Soviet Government must set the people in all its scope is – learn to work. The Taylor system, the last word of capitalism in this respect, like all capitalist progress, is a combination of the refined brutality of bourgeois exploitation and a number of the greatest scientific achievements in the field of analysing motions during work, the elimination of superfluous and awkward motions, the elaboration of correct methods of work, the introduction of the best system of accounting and control, etc . . . We must organise in Russia the study and teaching of the Taylor system and systematically try and adapt it to our own ends . . .
>
> Socialism is inconceivable without large-scale capitalist engineering based on the latest discoveries of modern science. It is inconceivable without planned state organisation, which keeps tens of millions of people to the strictest observance of a unified standard in production and distribution . . .[8]
>
> You all know that, following the military problems, those of defending the republic, we are now confronted with economic tasks. Communist society, as we know, cannot be built unless we restore industry and commerce, and that, not in the old way. They must be re-established on a modern basis, in accordance with the last word in science.[9]

Lenin's views, which reflected the sense of urgency caused by threats of invasion and economic collapse, were developed and extended by Stalin in the 1930s with the Stakhanovite (piece-work) movement.

While much has been written about labour/time-saving employment, scientific management, cost efficiency, labour theories of value and productivity, comparatively little research has been conducted into labour/time-absorbing employment until recently with the work of E. F. Schumacher, Ivan Illich and Kit Pedler.

PARKINSON'S LAW

C. Northcote Parkinson, a British historian, first proposed 'Parkinson's Law' in 1957 in an article in the *Economist*: 'Work expands so as to fill the time available to complete it.' A book entitled *Parkinson's Law* appeared in 1957, followed by *The Law and the Profits* (1960), *The Law of Delay* (1970) and *The Law* (1979). Parkinson, a conservative, is a passionate critic of bureaucratization who deplores the existence of Parkinson's Law. It does, however, recognize a profound truth – that service employment is inherently capable of mopping up far more workers than manufacturing because the services provided are consumed each day and then have to be provided all over again. (This applies to subsistence agriculture much as to British bureaucracy.) A manufactured product such as a car, a shirt or a bedpan is a consumer durable, a capital asset which is manufactured once in a comparatively short time but is subject to continual use – the car has to be driven, fuelled, washed and maintained, the shirt washed, ironed and mended, the bedpan used, emptied and sterilized.

Parkinson illustrates his law with what he calls 'our first and classic example of administrative proliferation', the Royal Navy and its civilian administrators in the Admiralty. In 1914 the Royal Navy had 542 vessels in commission and there were 4366 Admiralty officials and clerical staff (excluding officers and men, and dockyard workers). By 1967 there were 114 vessels in the Royal Navy, but the Admiralty staff had risen to 33 574. (If there were a comparable fall in the number of ships and a comparable rise in the Admiralty staff, we might see a navy of twenty-four ships with 258 184 public servants.) Parkinson also notes that between 1935, when Britain still had an empire, and 1960, when it did not, the number of administrators in departments historically linked with the empire had trebled.[10] Similarly, in 1945 the Australian defence forces had 564 448 members and there were 8173 public servants in the ministries of defence, navy, army and air. By 1978 there were 69 870 members of the Australian defence forces and 31 377 public servants in the Defence Department. (As an instructive comparison, as the international throughput of Royal Dutch Shell petrol has more than trebled its labour force has fallen by 36 per cent.) The point is made clearly in the diagrams on page 89.

Parkinson, having examined the rates of increased employment in British departments, concluded that his law could be stated in mathematical form:

In many public administrative departments not actually at war, the staff increase may be expected to follow this formula:

$$x = \frac{2k^m + l}{n}$$

where k is the number of staff seeking promotion through the appointment of

Royal Dutch Shell (International)

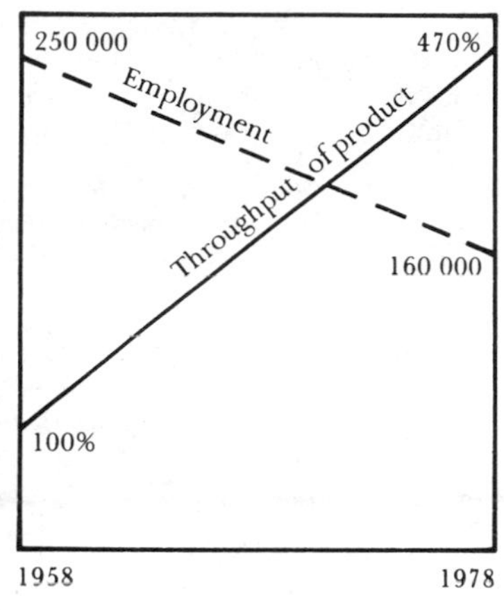

Australian government

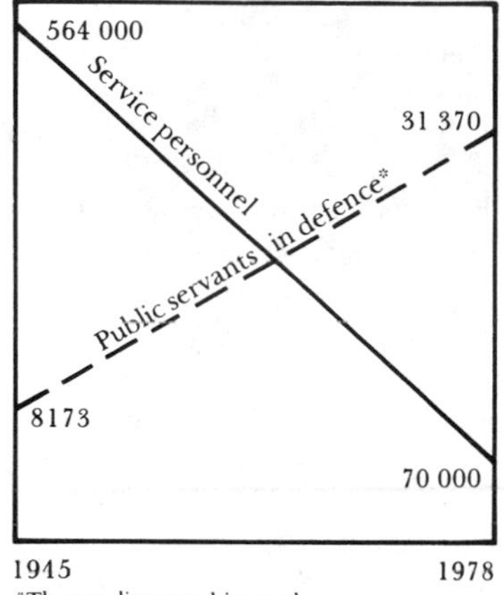

*The two lines on this graph are drawn to different scales.

subordinates; *l* represents the difference between the ages of appointment and retirements; *m* is the number of man hours devoted to answering minutes within the department; and *n* is the number of effective units being administered, *x* will be the number of new staff required next year.

Professor Dennis Gabor, in *Inventing the Future* (1963), saw Parkinsonianism as a natural defence mechanism by society against the psychological threat posed by technological advance and the unwelcome prospect of an age of leisure:

The present sum of working hours in the West, especially in the United States, is in no way in conformity with the level of our technology. It is kept up artificially, in the first place by enormous defence expenditures, and in the second place by waste. This is only partly a waste of products; to a much larger extent it is waste in unproductive man hours. The last is summed up in Parkinson's Law . . . The tool pushers who have become unnecessary change into paper pushers whose numbers can grow beyond any limit because they can always give work to one another. Yet I am inclined to consider this growth not as a tumour but as a healthy reaction of a virtuous society in which people have been brought up to work not only to earn money but also because they want to feel useful, and to keep their self respect.

Nobody has planned the growth of bureaucracy and of office work in general. It has come about by the unconscious wisdom of the social organism, which can be as admirable as the wisdom of the body. But who can be happy with a prosperity based on defence expenditures, waste, and Parkinson's Law? I consider it as dangerous, not because it is 'artificial' – all civilization is artificial

– but because it is unstable. It is a makeshift device which will collapse in all probability, and its collapse might lead to war, unless there is the vision to guide us into a stable state . . .

Symptoms such as material waste, Parkinson's Law, and irrational armaments can be interpreted in this sense as *defence mechanisms* of the social organism to stave off a danger – the danger of the Age of Leisure . . . It is not the symptoms which we must cure but their underlying cause – by bringing the fear out in the open and replacing it by hope in a worth-while future.[11]

Bertrand Russell anticipated Parkinson's Law in his important essay 'In Praise of Idleness' (1935) where he asked:

What is work? Work is of two kinds: first, altering the position of matter at or near the earth's surface relative to other such matter; second, telling other people to do so. The first kind is unpleasant and ill paid; the second is pleasant and highly paid. The second kind is capable of infinite extension: there are not only those who give orders, but those who give advice as to what orders should be given. Usually two opposite kinds of advice are given simultaneously by two organised bodies of men; this is called politics. The skill required for this kind of work is not knowledge of the subjects as to which advice is given, but knowledge of the art of persuasive speaking and writing, i.e. advertising.[12]

Waste and the absorption of time, work and capital

Waste is integral to defence expenditure – but better to waste than use in global or regional wars. In *World Military and Social Expenditures* (World Priorities Inc., 1979), Ruth Leger Sivard estimates that a total of $391 000 million was spent on war preparations in 1977. Governments see defence spending not only as an insurance premium but as a major employment factor. In 1977 there were 23 144 000 members of armed forces throughout the world – almost as many as there were teachers (27 700 000) and defence spending had enormous impact on manufacturing and many services, especially transport and research. World War II, the Cold War, the Korean War, the space race and the Vietnam War all stimulated many areas of the US economy. Massive unemployment has been avoided in the USSR by three factors: the size of its armed forces (nearly four million in 1980), employment in defence-based industries, and – as already mentioned – its enormous but relatively unproductive agricultural labour force. In Aldous Huxley's *Brave New World*, unemployment was avoided by keeping one-third of the population engaged in farming; while in Kurt Vonnegut Jr's *Player Piano*, those with redundant or non-existent skills were forced into the army or the 'Reconstruction and Reclamation Corps', known bitterly as the 'Reeks and Wrecks'. Michael Young's *The Rise of the Meritocracy* presents a future society in which the unskilled are cheerfully employed as domestic servants.

The surplus production of modern technological industry needs to be absorbed, recycled or distributed: it cannot be merely accumulated. For General Motors to have a stockpile of five million unwanted vehicles would be like having an army of five million corpses. The Great Depression was largely a result of accumulated surplus productivity. In *Monopoly Capital*, a Marxist analysis by Paul A. Baran and Paul M. Sweezy, an austere appendix by Joseph D. Phillips estimates the degree of surplus in peacetime as ranging between 40·4 per cent of GNP (1933) and 56·1 per cent (1963), but no attempt is made to calculate the impact on employment if the surplus had been cut back.

Many industries depend on what Thorstein Veblen called 'conspicuous consumption' and on 'built-in obsolescence', both of which generate material waste and also high employment. The most obvious example is the motor industry. As Jack Burnham wrote:

> When we buy an automobile we no longer buy an object in the old sense of the word, but instead we purchase a three-to-five year lease for participation in the state-recognised private transport system, a highway system, a traffic safety system, an industrial parts-replacement system, a costly insurance system, an outdoor advertising system, a state park recreation system, a drive in eating and entertainment system – and, not least of all, the general economic system.[13]

The motor car is central to our manufacturing and many service industries; increasingly, it dominates our lives, movement patterns, and urban development. Private motoring promises freedom of movement, provided that only a limited number try to use that right simultaneously. Within enclosed transport systems such as city road networks, however, mobility tends to be in inverse proportion to the total energy input at any given time. In *Energy and Equity*, Ivan Illich observed:

> The typical American male devotes more than 1600 hours a year to his car. He sits in it while it goes and while it stands idling. He parks it and searches for it. He earns the money to put down on it and to meet the monthly instalments. He works to pay for petrol, tolls, insurance, taxes and tickets. He spends four of his sixteen waking hours on the road or gathering his resources for it. And this figure does not take into account the time consumed by other activities dictated by transport: time spent in hospitals, traffic courts and garages; time spent watching automobile commercials or attending consumer education meetings to improve the quality of the next buy. The model American puts in 1600 hours to get 7500 miles: less than five miles per hour. In countries deprived of a transportation industry, people manage to do the same, walking wherever they want to go, and they allocate only three to eight per cent of their society's time budget to traffic instead of 28 per cent. What distinguishes the traffic in rich countries from the traffic in poor countries is not more mileage per hour of life-time for the majority, but more hours of compulsory consumption of high doses of energy, packaged and unequally distributed by the transportation industry.[14]

Australians, West Germans and Canadians are victims of auto-eroticism just as much as the Americans. Private motoring is one of the greatest factors of labour/time absorption in our lives, as we compete fiercely for the scarce resources of space, time and energy – not only against each other, but against energy-efficient time-saving public transport (when available). The car-dependent societies of this century, seen as a symptom of 'technological determinism', are considered in Chapter 10.

The concept of 'going to work' is a comparatively recent one: before the Industrial Revolution, people worked in or close to home, but with the development of public transport they were able to live further away from work. Now, with 70 per cent reliance on the car for metropolitan commuting (except to central business districts, where most commuters use public transport), the distance between home and work grows longer, the amounts of fuel and nervous energy consumed grow larger, and the total time taken from leaving home in the morning until returning at night often converts an eight-hour day to an eleven-hour day.[15] The declared aim of commuting by private transport is to save time: its actual result is to soak up time. Private commuting is also wasteful of resources, and destructive to the environment by both pollution and the impact of commuter freeways on inner-city areas. Nevertheless, the car is a major factor in promoting high growth rates of the GDP. In October 1979 the manufacture of motor vehicles and parts (including assembly) employed 86 000 people in Australia, while 163 500 worked in selling and servicing vehicles. It is notorious that car advertising concentrates on changes in styling which are frequent and arbitrary, involving variations in packaging and decoration rather than the basic design of engine and chassis (including safety factors). If the Australian community decided, in the interests of economy and efficiency, to squeeze one more year of life from existing vehicle stocks, to convert from the norm of the two-car family to the one-car family (or from one to none), or to rationalize journeys and make better use of time, space and energy, these decisions would all reduce employment. If Australian cities had adopted automatized transport grids such as the Personal Rapid Transit (PRT) system which operates on overhead guideways, this would have pursued the rational philosophy of 'moving people . . . not vehicles' but its effect on employment would have been adverse (except in the brief construction period). As Ralph Nader says: 'Every time there is an automobile accident the GDP goes up.'

A further illustration is the packaging industry – probably the worst example of a commercial activity completely devoted to waste. Raw material is converted to a finished product which is then imposed on the potential consumer only to become garbage – which is then destroyed, generating wasteful heat. Many commercial activities are *intended* to absorb time – the tobacco industry, liquor, hotels, clubs, tourism, much travel, gambling, the mass media, entertainment, racing, football. A utilitarian approach to life, eliminating activities which did not produce a quantifiable, tangible or durable result, would be very damaging for employment. Much service employment is inherently labour/time-absorbing and, as already suggested, the ratio between servers and served may be very low – one to one, or even less. A full-time chauffeur might only drive his employer for ten or fifteen hours per week, 48 weeks in the year. If he lost his job as a chauffeur and worked in motor manufacturing, he would produce the equivalent of one car every 5·3 weeks. (In Japan, he would produce the equivalent of two cars per week.)

We have come to accept much inherently wasteful economic activity without question, in the spirit of Hans Christian Andersen's story about the emperor's new clothes. We become conditioned to high energy usage and remain psychologically dependent as we move, in Ivan Illich's words,

> into the shrunk world of the powerfully rushed. The occasional chance to spend a few hours strapped into a high powered seat makes [the traveller] an accomplice in the distortion of human space, and prompts him to consent to the design of his country's geography around vehicles rather than around people.[16]

I propose Jones' Second Law, which states:

> *Employment absorption tends to be in an inverse relation to economic efficiency, to a chaos point of inefficiency beyond which labour is not absorbable.*

In other words: more waste = more jobs, but too much waste = no jobs. In the 1960s, an expert from Western Europe was invited to a country in Eastern Europe to give advice on the motor industry. He inspected the government-run motor plant and noticed that there were three or four workers performing tasks which would be carried out by one man in Western Europe. After the tour, the manager was bubbling with enthusiasm. 'This is the most efficient car plant in Eastern Europe. Isn't it the best factory you have ever seen?' 'Well, I . . . er, um,' the visitor responded uneasily. 'But don't you see?' said the manager. 'The factory is entirely appropriate to our needs. In this country we have a great need to create an industrial working class and we have very little need for cars. The factory fulfils both needs.'

Job creation as a form of labour/time absorption is not new. In the 1780s, Catherine the Great of Russia liked to impress foreign visitors with evidence of Russian development in the Crimea so her lover Prince Potemkin built cardboard model villages which, from a distance, achieved the desired effect. In France in 1848, General Eugène Cavaignac employed Parisians so that one gang would dig up ditches and they were then filled in by another. During Roosevelt's New Deal, the Works Progress Administration and other agencies sometimes employed many people in 'boondoogling' – the carrying out of pointless tasks to create work for work's sake.

Market and convivial economies

In *Tools for Conviviality* (1973), Ivan Illich examined the social problems caused by attempting to graft high technology on to a traditional social base, e.g. in Mexico and South America. He proposed the word convivial 'as a technical term to designate a modern society of responsibly limited tools' and defined a convivial society as one 'in which modern technologies serve politically inter-related individuals rather than managers'. Simon Nora, co-author of the French government report *L'informatisation de la société* (see Chapter 5), suggested adapting 'convivial' to refer to activities – e.g. libraries or schools – which are essentially intended to promote the quality of life rather than directly aimed at economic profit. I follow Nora's usage.

It will assist our understanding of changing patterns in employment to recognize that within all advanced economies there are two great economic systems, which co-exist but have completely different aims, organization, funding and technology: these may be called the market sector and the convivial sector.

The *market sector* is essentially capitalist (even where the state provides the capital) and investment aims to provide goods and tangible economic services – food, housing, clothing, energy, vehicles, entertainment, tourism – which are sold, with the intention of profit, to satisfy domestic needs (and, where appropriate, for export). The pricing of goods and services is determined by supply and generally holds sway, although it may be modified by demand for particular products. To survive and grow, the market sector must satisfy the test of cost-efficiency by making a return on capital invested. Economies of scale, mass production, and technological advances enable increasing volumes of goods and marketable services to be produced with a decreasing labour input. Some products are consumed (e.g. food, energy) while others (e.g. housing, clothing, vehicles) are durables. The values of this sector have changed very little from the time of Adam Smith and the classical school of economists. The market sector tends to be hierarchical, centralized and profit-based. Its products are easy to quantify because they are tangible in essence, and it can be seen where the sector pays its way (or does not).

The *convivial sector* is largely (and will be increasingly) publicly funded and/or managed. It includes education, municipal, health and welfare services, much of entertainment, sport and the arts, and some information services. Most of its products are non-economic and incapable of being exported or sold at a profit: many of them (e.g. nursing services) are 'consumed' each day and have to be provided again. The aim of this sector is primarily community well-being – something which economists and statisticians find notoriously hard to measure as yet – although it makes an indirect contribution to the market sector through education and health, and its influence on patterns of demand. The convivial sector is theoretically egalitarian and community-based, although in practice it may become a centralized bureaucracy.

The convivial sector is inevitably dependent on taxation gleaned from the market sector. The annual budget and permanent appropriations are set to meet the costs of the convivial sector, which in turn sets the level of market-sector tax rates – illustrating how the commercial economy has become subject to backdoor control through taxation irrespective of which party is in office.[17] (Of course, teachers, social workers and public servants pay taxes too, which reduces the funding needed.) It is a recurring complaint of conservatives that large sums can be pumped into the convivial sector with no marketable product emerging (e.g. education is seen as a cost to the community, not as an asset). It is often attacked for growing too rapidly and becoming too expensive, leading to popular demands for cuts in taxes and public spending – self-interested tax-payers who are young or middle-aged, healthy and employed, will resent supporting the aged, sick and unemployed.[18] However, elements of the market sector (such as motor manufacturing) are dependent on demand from convivial-sector workers to stave off collapse; and money given to welfare recipients does not stay in their hands – it is immediately transferred back to the market sector.

The market sector – agriculture, mining, manufacturing, construction and tangible economic services such as transport and utilities – has generally operated *à la* Adam Smith, and its employment levels are subject to a slow general decline. The convivial sector, essentially Parkinsonian in employment, may be particularly susceptible to future employment cuts, partly because this would be easy to effect technologically but also because this sector is often *perceived* as essentially parasitic – aiding the weak against the strong, crippling initiative through high taxation rates, and rewarding failure. Politically, how-

A two-sector economy

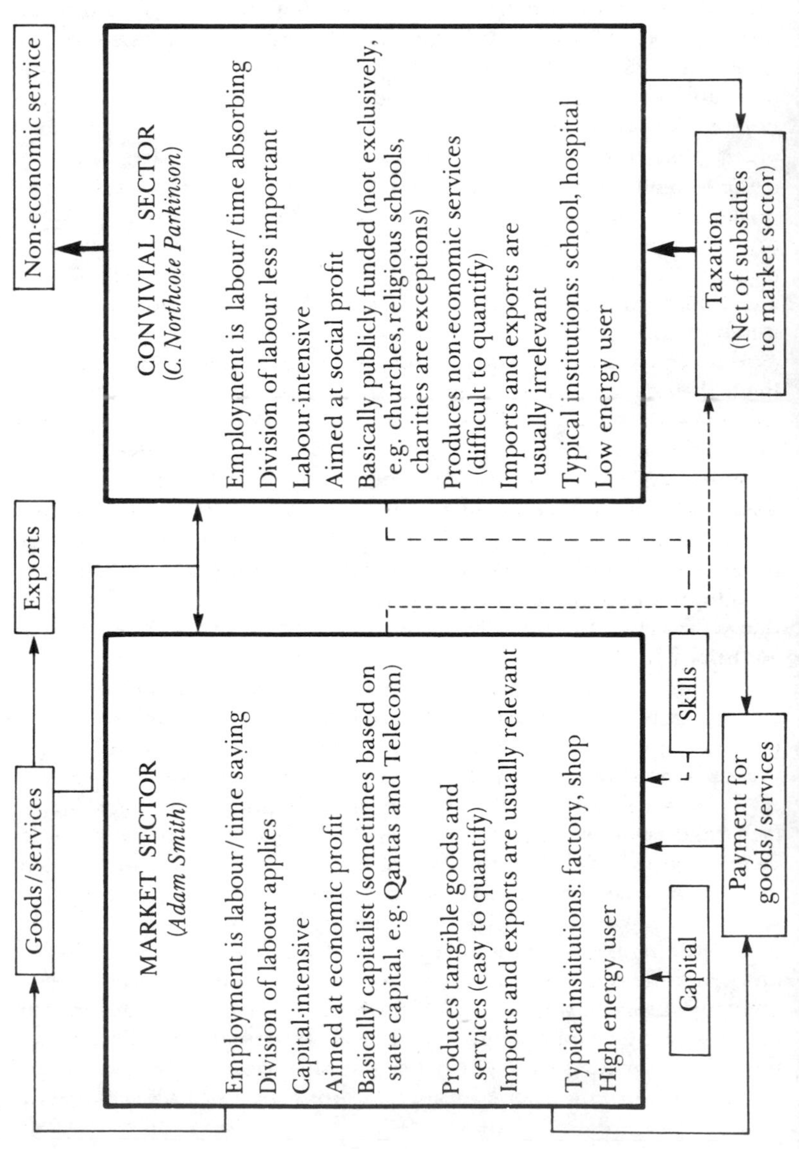

ever, both sectors should be seen as mutually dependent and we ought to reject the idea that only things which generate economic profit are worthwhile. There are good goods and bad goods, good services and bad services. Not even the most unreconstructed Friedmanite endorses every market activity, the heroin trade, or the pinball and skin-flick culture, all of which aim at economic profit and are not dependent on tax subsidy. Similarly, not all convivial activities are justified: there are grave dangers in over-centralizing bureaucratic empires which overvalue means and undervalue ends; some educational activity is trivial, and unworthy of time and effort by both teachers and the taught; small social investments made locally may produce greater results than large ones made to massive institutions. The cost-effectiveness test of the market sector needs to be applied to many convivial activities – and the social-benefit criteria of the convivial sector must be used in the market sector. Employment in both sectors should be encouraged, especially where it enriches and extends the diversity of human experience. It is better to be employed in a worthwhile market-sector job than to waste human resources in a worthless convivial one (like Dickens' Circumlocution Office in *Little Dorrit*) – and vice versa.

Adam Smith argued that individual self-interest was the factor that would best encourage economic growth (see Chapter 2). But Smith wrote at a time when population was determined by Malthusian forces – war, famine, disease – which were even more rigorous in their operations than the free market. The expectation of life was relatively low, with high infant mortality; cripples and invalids died at an early age; there was a high proportion of economically active people, and welfare services were only provided domestically. Now we live in a different world. Welfare is institutionalized and more need it. A major increase in longevity would dramatically change the proportionality of economically active and inactive people, and the convivial sector would grow very rapidly. The 'invisible hand' would be an increasingly inappropriate method of determining the allocation of national budgets.

At present, more people are employed for wages in the market sector than in the convivial sector, but the latter may have greater capacity for job creation. (If unpaid domestic workers are taken into account, the convivial sector is larger.) The most valuable single element in the market sector – mining – employs only 0·6 per cent of the labour force, mostly in Western Australia and Queensland (which see their assets as exclusive state property, causing some tension within the Federal system). Agriculture also employs few workers relative to its earning capacity. The dichotomy between the two sectors may assist us to identify future substantial job creation – but neither sector can do the impossible. The market sector will increasingly provide new jobs in areas of high efficiency only with heavy capital investment; the convivial sector cannot provide high levels of social productivity at low cost.

The two sectors are intended as broad illustrations only. The convivial sector is not synonymous with community services, as defined in the ASIC classification. There is also some ambiguity – many activities are in the market sector so far as the suppliers are concerned, but in the convivial sector from the viewpoint of the consumers (e.g. gambling, the liquor and tobacco industries, tourism and entertainment). The two sectors are not synonymous with the private/public or the profitable/non-profitable dichotomies either. Many public services – e.g. Qantas and Telecom – are clearly in the commercial area, operating in response to market forces. The provision of public transport in the various states is gradually moving from the market into the convivial sector, as are the more heavily protected manufacturing enterprises which are now virtually 'welfare industries'. Some, however, apply a more primitive analysis, in which the market sector is equated with virtue, growth and economic strength and the convivial sector (not so called) is identified with waste and irresponsibility.

THE VALUE OF THE INFORMAL ECONOMY

The convivial sector ought to include the informal economy of unpaid work, including domestic labour, DIY and charities. The present method of evaluating 'productivity' is far too narrow. The total output of the market sector is determined by the value of its products; the convivial sector's 'productivity' is measured essentially by adding the total wages paid – if no wages are paid, then no value is added. Sir Ieuan Maddock points out that

> it is a serious statistical aberration of the conventional run of national economic statistics that value added by individuals at home does not appear as production ... If all of us were to contract to do all the cooking, cleaning, gardening, painting, etc. for the house next door [and of course, if the house next door did it for us] then the 'national output' as measured by statisticians would be greatly increased without, of course, any increase of product.[19]

Hazel Henderson has argued

> for recognising the concept of household economics and the promotion of local self-reliance; to decentralizing economic power and exposing the absurdities of national economic statistics that do not recognise the enormous productivity of house-holds, simply because they are not, and probably never should be, a part of the cash-based economy ...

The increasing protest at the statistical blackout perpetrated for so long on the household economy is, of course, being spear-headed by women, who have been consistently ignored by economists' definitions of 'productivity' and 'value', as well as excluded from the GDP and their rightful access to retirement security. Many women fight these terrible injustices by going outside the home and competing successfully in the market economy, while others, together with concerned men, fight to restore the proper place of the family in our economic life and strengthen its role in the vital nurturing and socialising of the young and in maintaining intergeneration cohesion.[20]

In the long term, maintaining or increasing levels of employment depends on the following factors:

1 In the market sector, reversing economies of scale and providing assistance for low-volume high-employment production.

2 Using the convivial sector to raise levels of education and expectation so as to provide a greater demand for a variety of services (i.e. more emphasis on *range* than on *quantum*), many of which will be met from the market sector.

3 Recognizing that market and convivial sectors are absolutely interdependent.

5 Computers and Employment

The computer revolution . . . will alter the entire nervous system of social organisation.

Simon Nora

Computerization is the lead technology of the post-industrial revolution, and will help to create a post-service society marked by unprecedentedly rapid changes in the nature of work, society, communication and personal experience. A computer may be defined as a tool which converts data (raw material) into information (product) by following sets of programmes (instructions). It can operate without manual intervention, collate and store results, and is distinguished from other information-processing machines by its speed, capacity and versatility.[1] The work done by computers is often described as electronic data processing (EDP), or automatic data processing (ADP).

The physical units making up computer systems are collectively known as 'hardware'. They comprise the central processing unit or CPU (generally called main frame), storage device (memory), and input and output devices which feed material through the CPU into storage and retrieve it as required. The last three are known as 'peripheral units'. Inputs may be fed into the CPU by pre-recorded tapes, keyboards, cameras or light pens, and outputs may be 'read' on paper print-outs, charts, visual-display units (VDUs) or loudspeakers. The collective name for computer programmes is 'software' – and its production, often by computer manufacturers, is labour-intensive and expensive. It involves devising instructions for the computer to carry out specific tasks, such as evaluation of technical or commercial projects, costing a social welfare programme and issuing its cheques, planning optimum use of resources or updating files.

There are two types of computers – analogue and digital. Analogue machines are essentially problem-solving devices which convert physically variable factors (e.g. flow, speed, heat or pressure) into another physical variable, such as the movement of a needle on a calibrated dial. A conventional clock face, speedometer, voltmeter or a petrol gauge illustrate time, speed, energy or capacity by way of analogy. The great majority of computers are 'digital': they react specifically (or 'discretely') to exact inputs, convert them to numbers, process them, then convert the product to the desired form (often non-numerical). Any analogue signal form can be converted to digital form. The transmission of telephone speech will in the future be in digital form, with the analogue sound pattern being recreated at the receiving end. Music can be recorded in digital form, and the first commercial discs of this kind, hailed for their outstanding clarity in reproduction, were released in 1979. Visual images from body scanners (computers linked to X-ray machines) can be converted into numerical codes, recorded, stored and reproduced.[2]

In mathematical transactions, computers – and also electronic calculators – are programmed in decimal figures, convert them to binary notation, carry out the processing, then reconvert the binary result to decimal notation which can be read by the operator.[3] The long strings of binary digits which are required (1111101000 in binary notation = 1000 in decimal notation) are less complex than they seem. The computer has only to respond to an *0* or a *1*, one of two possible 'stable' states (sometimes known as flip-flops). This is achieved by almost inconceivably minute switching mechanisms which record impulses – an 'Off' signal means *0*, an 'On' signal means *1*. Transmission rates are now measured in bits, kilobits or megabits per second. The bit (*bi*nary digi*t*) is the smallest piece of information – the choice between alternatives (*0* or *1*, No or Yes, Off or On) – and kilobits and megabits are multiples of a thousand or a million. A byte is an arbitrary aggregation of bits, for example the binary equivalent of a complete word or number. Language can be encoded in any predetermined way, and several computer symbolic language codes have been devised for dealing with complex technical data, of which the best known are COBOL (*Co*mmon *B*usiness *O*riented *L*anguage) and two scientific systems, FORTRAN (*For*mula *Tran*slation), and ALGOL (*Algo*rithmic *L*anguage).

Computers can respond to any type of symbol – sounds, images, numbers, words – which is programmed for processing. They can, for example, read the universal product code (UPC) marked on goods at a supermarket, calculate the amount to be paid and make appropriate deductions from store inventories. The development of electronic and electrochemical sensors enables minute traces of biological or chemical pollution to be detected and recorded, or comprehensive meteorological maps to be made up from satellite signals. Computers can converse intelligently, conduct Rogerian psychotherapy, write sonnets, play chess, calculate orbits for satellites, resolve traffic congestion and help hopeful lovers find compatible partners. They can assist in design work through the use of 'graphic display': a designer using a light pen can draw an aircraft or car component and subject it to rigorous testing for wind-resistance, changing design configuration and wind-force strength as required. With the use of robotized output devices, computers can reject defective products on an assembly line, paint chairs, cut cloth or metal, lift heavy objects and operate where intense heat or absence of light and ventilation make work by humans impossible. Computers can process data at almost the speed of light and could add 1000 million seven-digit numbers in one minute. Laser memory systems, developed in the 1960s, can store 1000 000 000 000 bits of information (the equivalent of 4000 sets of the complete thirty volumes of *Encyclopedia Britannica*) and retrieve any piece of data within 8·6 seconds. Computers can make quantitative decisions – rejecting overweight or underweight merchandise – but can also be programmed to make qualitative judgments, for example in personnel management or job selection.

How computers evolved

Computers evolved over a long period as extensions of the abacus, the slide rule, and the adding machine, but operate with such astonishing speed and accuracy as to overturn the conventional wisdom about the relationship of productivity and wage rates. The abacus is the oldest digital calculating instrument (dating back at least 5000 years) and is still widely used in China, Japan and the Middle East for calculating and storing information.

John Napier, Laird of Merchiston (1550–1617) used marked ivory rods – the ancestors of moving parts on slide rules – as an essential part of calculating his logarithm tables (1617). Slide rules, in recognizable modern form, were produced independently in England by Edmund Gunter (1620), William Oughtred (1621) and Robert Bissaker (1654). Blaise Pascal (1623–62), French theologian, philosopher, mathematician, and author of the celebrated *Pensées*, built a digital calculating machine in 1642, using mechanical gears. It could add and subtract up to eight columns of figures, but not multiply or divide. Gottfried Wilhelm Leibniz (1646–1716), the great German philosopher, mathematician and rival of Isaac Newton, devised a calculator in 1671 called the 'stepped reckoner' which could – in theory – add, subtract, divide, multiply and extract square roots. In practice the machine, first exhibited in 1694, proved to be inaccurate and was abandoned. Leibniz did, however, make an enormous contribution to computing by pointing to the advantage of using binary notation.

From the sixteenth century, skilled craftsmen in Switzerland, France and Italy created 'automata' – clockwork performing models such as music boxes, operated by the rotation of cylinders and 'programmed' by the arrangement of tiny projections on the surface which set off specific responses (sound or movement) as they turned. René Descartes and Thomas Hobbes were both fascinated by 'automata' and often mention them in their writings. In 1725, a French inventor named Falcon devised a knitting machine 'programmed' by a perforated card; in the same year Basile Bouchon, whose father made automated organs, built a programmed silk loom in Lyon. His idea was developed by Jacques de Vaucanson (1709–82), who attained instant fame in Paris with his automated flute-player, tambourine-player and duck (which appeared to eat and defecate).

Joseph-Marie Jacquard (1752–1834), a weaver from Lyon, developed the work of Boucher and Vaucanson, and his automated loom (1801) used punched cards wrapped round a cylinder through which air was blown (like the paper rolls on a player-piano) to produce extremely elaborate and beautiful decorations of flowers and animals on silk and other fabrics. Jacquard looms were in use for more than 150 years. In 1820 Charles Xavier Thomas de Colmar, starting where Leibniz had left off, produced the first commercially successful calculating machine in Paris. This could add, subtract, multiply and divide, and remained on the market for nearly seventy years.

In 1834–35 Charles Babbage devised a plan for the first analytical computer – called the 'analytical engine' – which was to be fed by sets of Jacquard punch-cards to be read by mechanical 'feelers'. The machine anticipated modern computer techniques, e.g. 'conditional transfer' where the result of an intermediate calculation would automatically direct the machine to modify its own programme, and the use of a memory storage. As his first computer programmer, Babbage worked with Ada Augusta, Countess of Lovelace (1815–52), daughter of Lord Byron. Babbage's analytical engine was never completed, although he worked on it until 1854. It placed excessive demands on existing standards of engineering precision to cope with fifty counter-wheels in order to store 1000 numbers of fifty digits each, and the British government decided it was too expensive to fund any more.

George Boole (1815–64) an English mathematician and logician in Ireland, was a pioneer of modern symbolic logic. His 'algebra of logic' is one of the basic principles used in modern computer design, especially in 'binary switching', where quantities can be expressed by using only two symbols. It proposes that all sets of questions can be answered satisfactorily by 'Yes' or 'No' (for example, as in the game of 'Twenty Questions' where each 'Yes' or 'No' limits the range of future questions). Boole's 'binary logic' led to the creation of the NOT, AND and OR 'gates' in computers where contingent decisions result when one set of computations leads to another.

Herman Hollerith (1860–1929), an engineer who had taught at the Massachusetts Institute of Technology (MIT), combined Jacquard's punch-cards with electromagnetic sensors which could 'read' them automatically (1885). Using Hollerith's invention, the 1890 US census was tabulated in less than half of the time taken in 1880. Hollerith founded the Tabulating Machine Company in 1896. This merged with the International Time Recording Company of New York (1911), which later became IBM.

In 1898 Valdemar Poulsen invented a method of recording sounds (or coded images, although this was not then recognized) on magnetized wire. Not commercially developed until 1927, Poulsen's recorder was an essential element in the first electrical analogue computer. This was a differential analyzer which could handle up to eighteen independent variables, designed in 1930 by Vannevar Bush (1890–1974), an American engineer working at MIT.

The formal theory of universal computer operations was devised independently in the 1930s by Alan Turing and Alonzo Church.[4] In 1936 Turing described how a universal computer would work, where encoded storage materials such as tapes or discs could be shuttled backwards and forwards, reading, calculating and altering as programmed. He invented an abstract model of a computer (often called a 'universal Turing machine') which embodied the logic of any computing machine, both past *and* future. Britain, however, ultimately lost the computer race to the United States.

FOUR GENERATIONS OF COMPUTERS

World War II and the beginning of the Third Industrial Revolution in 1942 stimulated the use and development of large-capacity calculating devices to solve problems in logistics, cryptography and anti-aircraft prediction. Harvard Mark I was an electro-mechanical calculator using decimal notation, built between 1937 and 1944 by Howard Aiken of Harvard University and IBM engineers. (It was not a computer as currently defined because it lacked a stored programme which could alter itself.) This was followed by Electronic Numerical Integrator and Computer (ENIAC), the first completely electronic computer, built by J. Presper Eckert Jr, John W. Mauchly and J. G.

Brainerd at the University of Pennsylvania and completed in 1946. Electronic Discrete Variable Automatic Computer (EDVAC), in 1947, represented a giant step forward. Devised by John von Neumann, this machine was essentially ENIAC plus binary notation and stored internal programming.[5] 'First-generation' computers depended on thousands of thermionic valves (called vacuum tubes in the US) which were complex and expensive to make, occupied a relatively large space, required high-voltage inputs, and generated considerable heat which reduced capacity and caused operational failures – i.e. they were huge, expensive and liable to malfunction. Australia had one of the earliest first-generation computers, CSIRAC, designed in 1947 by Maston Beard and Trevor Pearcey: it was fully operational from 1951 to 1964.

'Second-generation' computers were made possible by the invention of transistors, small 'solid-state' semiconductors which could act as switching devices or amplifiers, replacing thermionic valves at a fraction of cost, size, energy use and heat generation. (Solid-state devices are units made of semiconducting material – such as germanium, silicon or the 'crystal' of early radio sets [galena] linked to electrodes. The input signal disturbs relationships between atoms arranged in a 'lattice' formation inside the semiconductor, performing the same function as the thermionic valve.) In 1947–48 the transistor was invented at the Bell Telephone Laboratories by William B. Shockley, John Bardeen and Walter H. Brattain. In 1956, they shared the Nobel Prize for Physics. Transistor development in the 1950s permitted a dramatic reduction in computer size and cost, together with a vast increase in capacity. (By the 1960s transistors had diminished in size a thousand-fold.) There were also many improvements in computer storage and retrieval capacity in the 1950s.

'Third-generation' computers resulted from the development of integrated circuitry in a single manufacturing process. Before 1960, most electronic devices packed an average of one component into each square centimetre: with the first integrated circuits in 1962, this increased to six per square centimetre. Medium-scale integrated circuits (MSI) raised this to seventy by 1965 and large-scale integrated circuits (LSI) to 6500 components per square centimetre by 1969. Very large-scale integrated circuits (VLSI) contained up to 200 000 components per square centimetre by 1979.

Fourth-generation computers based on microtechnology date from 1970–71. 'Microchips' are thin wafers of silicon less than a centimetre square, on which all components are etched by a combined photo-lithographic and chemical process (or by laser in the latest models), containing thousands of microcircuits. The most common chips are sold for between twenty and thirty cents each, and it was this low cost which made miniaturization and microminiaturization possible.

Microchips were developed at Stanford University, California, by Robert Noyce, a PhD from MIT, and other members of a team originally recruited by William Shockley. Eight team members broke away from Shockley and set up several research and manufacturing plants – of which the Fairchild Camera Corporation (1957) and the Intel Corporation (1968) are the best known – in the Santa Clara Valley south-east of San Francisco (often called Silicon Valley). The United States has about 67 per cent of the world's microchip market, Japan 15 per cent, Europe 10 per cent, and the USSR 8 per cent. Texas Instruments of Dallas is the largest producer, followed by Philips, IBM, Intel, National Semiconductor, Motorola and Hitachi. Many US chip-makers send their products to Indonesia, the Philippines and Singapore to have the leads attached – a relatively labour-intensive activity. In 1971, M. E. Hoff Jr of the Intel Corporation invented the micro-processor (MPU), a 'computer-on-a-chip' comprising a miniaturized logic and calculating faculty containing 2250 microminiaturized transistors. Its central processor is smaller than the typed initials MPU and works twenty times faster than ENIAC.

Magnetic bubble memories were invented at the Bell Laboratories in 1966. They consist of tiny magnetic bubbles or 'domains' which move about within crystal chips (such as garnet) under the influence of magnetic and electrical fields. Five million bubbles can be stored on a chip about 1·5 centimetres square, and it is hoped to achieve a capacity of fifteen megabits per square centimetre: 'Someday', Bell claim in advertisements, 'a bubble chip the size of a postage stamp may store the contents of an entire telephone directory.' Since 1966, the size of bubble devices has been cut by a third, operating speeds are ten times faster, and four times as many bubbles can be placed on a single chip.

By 1975 computers such as ILLIAC IV and STAR 100 could perform 100 million operations per second. Theoretically, the speed of computer operations is only limited by the speed of light (300 000 kilometres per second) but internal resistance, such as heat generation, slows the process considerably. As computers become more compact, the problem of heat dispersal is greater and some sophisticated computers incorporate freon refrigerants. In 1968 Brian David Josephson, Professor of Physics at Cambridge University, designed a configuration called the 'Josephson junction' – a device consisting of a thin oxide-insulating layer placed between two metals, which enabled low-temperature high-speed micro-miniaturization (it permits a switching speed thousands of times faster than the microchip).[6] In 1973 the Nobel Prize for Physics was shared by Leo Esaki, Ivar Giaever and Josephson for their investigation of tunnelling in semiconductors and superconductors.

Miniaturization

Micro-electronics permits an exponential rise in output together with an exponential fall in total inputs – energy, labour, capital, space and time. In economic history there is no remote equivalent to this.

The first electronic computer, ENIAC, cost $US5 million to build (about $20 million in 1980 currency), occupied a space of 90 cubic metres, (9m × 5m × 2m), weighed 30 tonnes, used 18 000 vacuum tubes – of which some hundreds had to be replaced each day – consumed as much power as a locomotive (about 140 kilowatts), and generated considerable heat. A modern micro-computer costs $2000, is 1500 times smaller, 10 000 times cheaper, 17 000 times lighter, with a 'mean time between failures' measured in years, uses 2800 times less power, and generates very little heat. The micro-computer is forty times faster than ENIAC, and its memory capacity is 400 times greater. It is difficult to think of an appropriate analogy to illustrate the significance of the miniaturized revolution: it is as if modern aviation had begun with the jumbo jet and evolved towards an aircraft which was cheaper, lighter, faster and safer than the Wright Brothers' model at Kitty Hawk. In Britain, the somewhat sour story is told of the micro-electronics firm which grew so rapidly that it had to move into smaller premises.

Computers do not yet have the flexibility of the human brain, although speed, accuracy and storage capacity are far greater. (An IBM 3850 can store on-line to a computer up to 472 000 million characters of information – as many words contained in 27 million pages of a typical daily newspaper). In the 1960s, a computer with as many nerve cells as a human brain would have occupied a space as large as London's Albert Hall: by the end of the 1970s it would have been reduced to the size of a large suitcase. Since human brain cells are scarcely bigger than the bubbles in magnetic memories, it is likely that microminiaturization will permit a computer to be built which is equal in capacity and size to the human brain.

We have all grown up with maxims like 'You can't have your cake and eat it too' or 'A rolling stone gathers no moss' – the folk wisdom that for every advantage we obtain there is a corresponding disadvantage or price to be paid. If we want a good seat at the opera or football, or a house in a good location, we expect to pay more for it.

To expect a greater return for a smaller investment in time, money and effort seems contrary to every conventional precept. If we dig a trench with a teaspoon, the effort required for each quantity of soil removed would be very small – but the time taken to complete the task would seem endless. If we use a spade, far more effort is put into every movement but the task is achieved much more quickly. But imagine that we could dig the trench with no more effort than is needed to wield a teaspoon but in far less time than would be taken with a spade. This is precisely what miniaturization does: we are able to maximize two functions at one and obtain greater capacity at lower cost. Of course, ever since the Industrial Revolution began, there have been falls in the comparative costs of manufactured goods – but these falls were not measured in hundreds or thousands of times the unit cost. IBM asserts that a set of computations costing $1·26 in 1952 costs only 0·7 of a cent in 1978 – reduced by a factor of 180, or by 400 allowing for inflation (*New Yorker*, 19 June 1978). As technology becomes more sophisticated it is relatively cheaper. The cost of on-line storage was reduced from $25 per million characters in 1965 to 50 cents in 1974.

Watt's steam engine forced many work-horses into premature retirement – but it created jobs for the men who made the machines, operated them, maintained them, built the factories to house them, dug up the coal and supplied the water which ran them. But suppose that one of Watt's contemporaries had produced a steam engine of equal capacity which required little maintenance or supervision, virtually no fuel, could be bought cheaply, and was small enough to be slipped into a coat pocket: what impact would that have had on work demand? Imagine a cow which gave abundant milk twenty-four hours each day, was not dependent on good pastures, kept itself clean and could be carried around in a handbag: would that not have revolutionized agricultural employment?

In 1950 the father of cybernetics, Norbert Wiener, in his book *The Human Use of Human Beings: Cybernetics and Human Beings* wrote these prophetic words: 'Let us remember that the automatic machine . . . is the precise economic equivalent of slave labor. Any labor which competes with slave labor must accept the economic consequence of slave labor.'[7] The labour-displacing machines *can* work twenty-four hours a day. They do not require lighting, heating and ventilation in their workplaces. They do not need annual holidays, public holidays or weekends. They do not receive overtime or penalty rates. They do as they are told and never argue back. They do not go on strike and they do not join unions. They are not eligible for long-service leave, sick leave or workers' compensation. They have no meal breaks. They do not seek wage rises and they make no demands for industrial democracy.

In the present distribution of wealth and power in Australia, the new technology is often used as an instrument of the strong against the weak and the rich against the poor. This leads to a strengthening of the skilled and accomplished against the unskilled or deskilled. It is all happening with only the barest expression of interest by Parliament, and with little public understanding about what is going on.

Some government responses to technological unemployment

In the 1950s and early 1960s, there was premature and inappropriate anxiety in the United States and Western Europe about job displacement by automation. At that time cybernetic techniques were crude, expensive and accident-prone, service employment was growing rapidly, and so many basic needs were unsatisfied that some European nations (notably West Germany) imported 'guest workers'. President Lyndon B. Johnson appointed a national commission to examine the employment effects of computerized labour displacement, and in 1966 the commission reported:

> Our study of the evidence has impressed us with the inadequacy of the basis for any sweeping pronouncements about the speed of scientific and technological progress . . . Our broad conclusion is that the pace of technological change has increased in recent decades and may increase in the future, but a sharp break in the continuity of technical progress has not occurred, nor (since most major technological discoveries which will have a significant economic impact within the next decade are already in a readily identifiable stage of commercial development) is it likely to occur in the next decade.[8]

This was at the highpoint of Keynesian orthodoxy and confidence. The commission assumed that although automation and the growth of capital intensity would create structural unemployment, an essentially perfect labour market would absorb workers elsewhere with little disruption due to continuous economic growth and population increase. The significance of micro-electronics in the 1970s was not envisaged by the commission, whose findings were obsolete less than a decade after they were written. (However, the report was sometimes cited as if it were the last word on technological unemployment.) Fifteen years later, despite development of low-cost, highly efficient miniaturization and robotization, there remains in some areas an equally inappropriate optimism about employment prospects in the

1980s. Policy responses which evolved in a period of labour scarcity are being applied – without success – to a period of surplus labour.

In Great Britain the most quoted examination of the technology/employment question was an official paper by the Central Policy Review Staff, entitled 'Social and Economic Implications of Microelectronics', an anonymous document of twenty-four pages published in November 1978. The only statistics in the report appear in a section entitled 'Computers in the Public Service'. They deserve to be reproduced in full:

> [Para. 16] . . . By 1977/78 some 1650 (mainly clerical) posts had been saved (partly offset by some 450 computer related posts). This picture is repeated in other computer installations, but taking a wider view there has between 1970 and 1977 been a growth from 170,000 to 200,000 staff in categories most likely to have been affected by computer installations.
>
> [Para. 17] . . . In 1977 the numbers of staff actually engaged in computer operations was about 14,000 and the number 'freed' by computers was several times that figure.[9]

No statistics are provided for the impact of micro-electronics on manufacturing or service employment; there are dangers in trying to apply evidence based on civil-service employment to the market sector. Para. 13 reads: '. . . the real employment gains will accrue to those countries which can translate microelectronic innovations into new, attractive, inexpensive products for mass consumption.' It is obvious that the authors had no idea what these products would be – nor who would buy them. The only example given was 'the incorporation of semiconductors into watches' – a particularly maladroit selection if it was hoped to identify an example of job expansion.

The British Department of Employment set up a manpower study group in July 1978, and in December 1979 its members (Jonathan Sleigh, Brian Boatwright, Peter Irwin and Roger Stanyon) published a 110-page report, 'The Manpower Implications of Micro-Electronic Technology'. They concluded that:

> Britain has no option but to seek to adapt to micro-electronic technology at least as fast as our competitors. This formulation may represent something of an oversimplification. What lies behind it is the belief:
>
> > that Britain's industrial base has recently shrunk and is in great danger of shrinking further for reasons of competitive failure; and that micro-electronic technology could be of significant assistance to us in halting and perhaps even reversing the trend.

They argued that human factors such as the attitudes of management and labour will prevent micro-electronics being introduced at too rapid a pace, and endorsed the proposals of the Trades Union Congress (TUC) for new technology agreements which 'offer a constructive approach which largely reflects best practice in Britain'. The authors thought that the Japanese situation offers 'some clear examples of how maximum exploitation of new technology can be combined with reasonable guarantees of continuing employment'. The Sleigh Report examines a number of industries for specific growth opportunities and concludes that electronics has the best prospects. (Nevertheless, output would need to grow by 15 per cent per annum to maintain existing employment in electronics.) It identifies many areas where skilled workers such as technicians and engineers are needed, but the unskilled are never mentioned at any point.

In Australia, the Minister for Employment and Youth Affairs (Ian Viner) confirmed in answer to a question on notice his inability to provide any specific examples of research into the impact of computerization on unskilled or semi-skilled workers.[10] The answer cited twelve studies of specific industries carried out before 1972 and some later studies on EDP, but made it clear that the larger issues had never been examined. The most serious government investigation of computers and their employment effects was carried out in France.

THE NORA REPORT

In December 1976, the then President Valéry Giscard d'Estaing commissioned a report on the impact of computers on employment in France from Simon Nora, Inspecteur-Général des Finances.[11] *L'informatisation de la société*, by Simon Nora and Alain Minc, was published in May 1978, became an immediate best-seller, and is the most comprehensive and convincing treatment of this complex subject. The report comprises one volume of argument and four of appendices. The first volume was translated into English, German, Italian and Japanese. The English version was published by the MIT Press as *The Computerisation of Society* (1980). The Nora Report proved extremely unwelcome to the French government because it cast doubt on the economic viability of the 'seventh national plan' which anticipated an increase of 215 000 jobs in manufacturing and 1 335 000 in the service sector in the next decade. The work was largely ignored in the English-speaking world, partly because of the language barrier but also because its conclusions are unwelcome to technological determinists.

Nora and Minc coined the word *télématique* – telematics – to describe the synergistic relationship between telecommunications and computers. Under the heading 'No More New Jobs in the Service Sector', they wrote:

With telematics, the service sector will in the coming years undergo a jump in productivity comparable to the gains in productivity enjoyed by agriculture and industry in the past twenty years.

Although it is not possible to make a thorough evaluation or fix the tempo of this development, an examination of several large sectors will show its importance.

1. In *banks*, the installation of new computer systems would permit employment reductions affecting up to 30 percent of the personnel over ten years, but this does not mean that workers would have to be discharged. In effect, these reductions are a measure of the numbers of additional personnel that would be required under current rates of productivity to meet the coming demand, and telematics would make additional hiring unnecessary . . .

In other words, the 30 percent savings in jobs is not the automatic result of a transformation of the computer system. Even if management wanted it, this policy would run up against structural red tape, individual resistance, and pressure from the unions. There is no room for inertia, however, since competition will force banks that may be tempted to assume a passive role to keep up with the more dynamic domestic banks, and even more with their foreign counterparts.

2. In *insurance*, the phenomenon is even more pronounced. Job savings of approximately 30 percent are now possible within ten years. Some companies, fearful of the reaction from their personnel, have put a moratorium on the installation of telematics systems. Once again, however, it cannot be delayed indefinitely, since the freedom to establish insurance companies within the EEC will introduce foreign competition.

3. In *Social Security*, the movement will be slower, since data processing is still traditional in form, with large centers and massive, cumbersome processing. Even if no outside pressure acts to shake the inertia characterizing its organizations, its traditions and its regulations, then the need to keep down costs will eventually do so. It is difficult to say how long this will take, however, when causes are the same and effects are the same. The likelihood of achieving job savings through telematics will be the deciding factor.

4. For the *postal services*, the foreseeable reduction in manpower is the result of another type of competition. The new computerization will not bring about massive gains in productivity in this type of work, but the rapid development of telecopying and teleprinting, soon to become a reality, and the longer-term prospect of home newspaper publishing are all factors working in favor of a decrease in postal activity . . . There is little doubt, however, as to the inevitability of the substitution of telecommunications services for postal services and the resulting effects on employment.

5. The computerization of *office activities* will affect the 800,000 secretaries in this huge sector spread out over the entire economy. The development of data-processing networks, telecopying, and the incorporation of microprocessors into typewriters are leading to a new type of secretarial pool, one more involved in supervision than in the performance of tasks . . .

Thus we have five dissimilar service functions – banks, insurance, Social Security, postal and office work – with greater or lesser degrees of computerization, whose effects are sometimes direct and sometimes associated with changes in the volume of traffic, and which operate under constraints that in some cases are the result of foreign competition and in others induced by political pressures to reduce costs. Despite these differences, the conclusion is the same for all: within the next ten years, computerization will result in considerable manpower reductions in the large service organizations.

Can this conclusion be extrapolated over the entire tertiary sector? Intuition says yes, but its extent cannot yet be measured, at least not on the basis of the few projections made in the preparation of this report.

The change in computer technology will be accompanied by more rapid automation of industrial enterprises. It will affect internal 'tertiary' activities as well as production and will involve robotics as well as automated systems.[12]

A report by Siemens A. G. of West Germany, entitled 'The Office in 1990', also appeared in 1978. As one of Europe's largest suppliers of office equipment, Siemens approached the question of job displacement from a different starting point to that of Nora and Minc but reached identical conclusions about the extent of lost employment.

Computers: the question of positional advantage

Owning a computer undoubtedly confers competitive or 'positional' advantages unless, or until, competitors acquire computers too.[13] A person who stands on a butter-box to view a procession will secure a positional advantage over other people in the crowd – unless everybody else uses a butter-box, in which case the status quo is restored. The first grocery in town to use a motor truck for deliveries would have had a distinct advantage over its rivals and the purchase could be seen as evidence of imagination, resourcefulness and capacity for innovation. The grocer may have gone from strength to strength, increasing his workers by half and doubling his turnover: nevertheless, if his rivals went under total grocery employment in town would fall, and the new technology of that time would have been a major factor. A supermarket which puts three family stores out of business might increase its labour force but still cause a net loss in local employment.

The tendency towards consolidation and oligopoly continues in Australia, with record profits (and some increases in the labour force) in large and often overseas-controlled enterprises while many small locally owned firms collapse into bankruptcy. Two contrary movements may occur simultaneously: the large firms computerize and grow, while the small ones fail to computerize and go under. The overall result is a net job loss – but the computer-users maintain their positional advantage where, for example, employment goes up by 10 per cent and output increases by 80 per cent. The computer is a vital element in a winning strategy – a tool for determining who wins and who loses. But winning on a large scale involves a major positional advantage – most must lose that some may win: the positions are entirely relative.

Managements adopt computer technology in order to gain economic advantage, but they do not always understand how to use the computers to capacity. Dr Barry S. Thornton and Philip M. Stanley concluded that 11 000 computers had been installed in Australia by 1978, most of them in the smaller categories:

Almost 24,000 persons are employed directly in the skilled areas of EDP and a further 53,000, at least, are employed in data input preparation, data control and as operators . . . The manpower equivalent of the computers installed in Australia is almost 3 million people. Of course, nothing like this number have been displaced. However, to be worthwhile from an economic investment point of view, the huge investment made so far (nearly $1,000 million per year in hardware, salaries, software, supplies and ancillary staff) would have had to be able to displace some 200,000 jobs to have been worth doing at all. Very few statistics exist on the number of jobs consumed by computers. As a conservative estimate – after allowing for 53,000 people involved in data entry and operating – there would be in the region of 150,000 low-grade clerical jobs abolished as the net result of computerisation. At the good end, 24,000 jobs have been created for the highly skilled, in systems and programming.[14]

In an updating of this report in June 1980, the Foundation for Australian Resources concluded that Australia was still about 50 per cent under-computerized but that, on a conservative estimate over a twenty-year period, 244 000 jobs had been eliminated by computers: after allowing for 77 000 jobs created in computer and related industries, a net loss of 167 000 jobs. Computers are *intended* to displace labour. Their manufacturers promise to reduce manpower, and there is no reason to doubt them. It would be the greatest fraud since the South Sea Bubble if managements were persuaded to invest in computers in order to cut wage bills relative to output, and then found that computers actually increased employment relative to output. Computerization has the same aim as contraception – to eliminate people. The contraceptive pill has done for the birth rate what the silicon chip is intended to do for the labour force. There are three alternative scenarios:

1 Smaller labour input + computer = same output.
2 Same labour input + computer = larger output.
3 Larger labour input + computer = very much larger output.

An American economist, Herbert Grosch, has devised a method for calculating job displacement by computers.[15] A Grosch shift unit (GSU) is the amount of work which a $US50 000 computer will carry out in one shift: thus, as computers fall in cost, the GSU rating will rise. One GSU equals 15 manpower units, so that the capacity of an IBM 360/20 would be rated at 60 manpower units, i.e. 4 GSU. Grosch's Law states that the computing power of a central processor increases as the square of its price, i.e. that a $40 000 computer will have four times the capacity of a $20 000 unit.

Is computing the 'X' industry?

Will computing emerge as a major employer – including designers, manufacturers, salespeople, maintenance workers, and operators – comparable to the motor industry or civil aviation? It seems unlikely, for as production volume increases, economies of scale will lead to a relative decline in the numbers of people employed (even if there are small increases in absolute numbers).

IBM, the world's largest computer manufacturer, has a relatively modest labour force. *Computer World* (May 1978) estimated the market shares of installed general-purpose computer equipment – that is, excluding military command and control, industrial process control or other 'dedicated' single-purpose machines (e.g. for medical or laboratory use) – throughout the world as follows:

IBM	58·4%	CDC	2·9%
Honeywell	7·0%	ICL	2·5%
Univac	6·4%	Siemens	2·1%
All Japanese	5·2%	NCR	1·8%
Burroughs	4·8%	Others	8·9%

In *Datamation* (December 1978), Rolf Emmett wrote: 'Today IBM has a backlog of more than four times the computing power it has ever shipped. Or, put another way, IBM's backlog is greater than the total computing capacity ever used on this planet.' In total revenue from all computing (including software), the ranking was:

IBM	34·4%	Univac	3·6%
All Japanese	10·8%	DEC	2·9%
Burroughs	4·2%	Honeywell	2·6%
NCR	3·9%	CII-HB	2·1%
CDC	3·7%	Others	31·8%

In 1978, IBM employed 325 000 people throughout the world to produce $US21 076 million in goods and services – and the company payroll includes sales, promotion, research and the production of software. IBM Australia Ltd, the market leader, employs 2700 people in all categories (about 0·8 per cent of the corporation's worldwide labour force). There appear to be no published figures for total employment in the world's computer suppliers, but estimating from IBM's employees, from 600 000 to a million people would be the appropriate range.

A vague but optimistic guess about Australian employment in computing was presented to CITCA by the Australian Computer Equipment Suppliers' Association:

Our estimates of people directly employed in the computer industry fall into three categories:

- i The total number of employees of computer supply companies is estimated to be between 12,000 and 14,000;
- ii We estimate that Systems Houses and Original Equipment Manufacturers employ a similar number of people as computer suppliers, i.e. 12,000 to 14,000;
- iii The number of people who are directly employed in the computer departments of computer user organisations is estimated to be between 45,000 and 50,000.
- iv This gives a total of between 69,000 and 78,000 people directly employed in our industry.[16]

General Motors employed 839 000 people in 1978 – and its share of the car industry is far smaller, pro rata, than IBM's in computers. The GM figure excludes sales and service personnel (unlike the IBM figure given above). The US motor industry, both directly and indirectly, accounts for one job in every six. Computing is a very modest employer by comparison: the car is a paradigm of labour-complementing technology, and the computer one of labour-displacing technology.

Distinguishing between labour-complementing and labour-displacing technology

Car	*Computer*
Major emphasis on personal or family use	Major emphasis is on industrial and institutional use
Typically, each has its own operator who owns his/her own vehicle	Typically, a shared facility and operators do not own computer
Each car has a very long 'downtime' (i.e. it tends to be used for an average of 2–3 hours daily)	Major computers have very short downtime – and the extreme portability of information transactions means that computing can be done anywhere, at any time
Generates large employment in: storage waste disposal fuel supply repair and maintenace	Generates relatively small employment in: printout paper tapes preparing software
Requires expensive road systems which are labour-intensive to build and maintain	Depends on wired networks which are not labour-intensive to maintain
Often violent and/or dangerous in operation	Never violent or dangerous in operation

Car	*Computer*
Contributes to employment in: police, hospitals, undertakers, florists, insurance, motor sports, hire purchase	No analogous employment generated
Large employment in manufacturing and sales	Relatively small-scale employment in sales (no significant manufacturing in Australia)
Operating speed directly related to allocations of space	Operating speed may be inversely related to space requirements
Prices rise relative to capacity	Prices fall relative to capacity

Even with a computer in every household – as seems probable in the future – it is unlikely that much more employment would be generated than in the existing television industry. 'Computocentric' households may have significantly less demand for live entertainment, eating out, books and newspapers, public transport and other labour-intensive activities. In addition, it may lead to an even higher degree of machine dependence as people, like Mr Chance in the 1979 film *Being There*, prefer to live vicariously through electronic imagery than to respond to personal contact in a broader social context. There is also the question of how semi-skilled and unskilled workers can compete economically with the new technology. If the gap between the cost of labour and that of technology is growing wider, why would industry and commerce choose to create jobs when they can increase profit by decreasing labour costs and using the new technology?

In addition, computerization makes it possible to eliminate not only routine and repetitive work, but also some work requiring lesser intellectual skills due to the development of 'artificial intelligence' (AI).

ARTIFICIAL INTELLIGENCE

Exponential increases in capacity to collect, store and disseminate data have led to the development of information theory. Claude E. Shannon, formerly of the Bell Laboratories and later Professor of Computer Science at MIT, is regarded as having originated information theory with William Weaver in their paper 'The Mathematical Theory of Communication' (1949). He argued that information flow should be treated like a physical quality such as matter or energy, and worked for many years attempting to find whether there is an upper limit to the amount of information any given channel may transmit. The question is still open.

Herbert A. Simon and Alan Newell have worked on the development of artificial intelligence and general problem-solving (GPS) – which seem to proceed on the premise that there is greater economy and efficiency in trying to educate human beings to react to stimuli in the same way that a computer responds to programming, than in fitting the computer to fulfil human needs. In 1972 they published *Human Problem Solving*. Dr Simon, who is Professor of Computer Science and Psychology at Carnegie-Mellon University in Pittsburgh, was awarded the Nobel Prize for Economic Science in 1978 – the first computer scientist so honoured.

In his important book *Computer Power and Human Reason: From Judgment to Calculation*, Joseph Weizenbaum, Professor of Computer Science at MIT, expresses reservations about the evolution of a 'computocentric' approach to problem-solving. He gives the example of the ELIZA programme, which he created to illustrate how a computer could question and respond to patients in the same way that a psychotherapist would. Weizenbaum says that he intended ELIZA as a parody, only to find that it was praised as a valuable and time-saving form of therapy. He is concerned by the implications of the equation 'humans = information processing systems = computers', which he sees as an ominous parallel to B. F. Skinner's dictum that 'A scientific analysis of behaviour must . . . assume that a person's behaviour is controlled by his genetic and environmental histories rather than by the person himself as an initiating, creative agent.' Weizenbaum writes of artificial intelligence:

> the goal of AI is to understand how an organism handles a range of problems co-existensive with the range to which the human mind has been applied. Since the human mind has applied itself to, for example, problems of aesthetics involving touch, taste, vision, and hearing, AI will have to build machines that can feel, taste, see, and hear. Since the future in which machine thinking will range as widely as Simon and Newell claim it will is, at this writing, merely 'visible' but not yet here, it is perhaps too early to speculate what sort of equipment machines will have to have in order to think about such human concerns as, say, disappointment in adolescent love. But there are machines today, principally at MIT, at Stanford University, and at the Stanford Research Institute, that have arms and hands whose movements are observed and co-ordinated by computer-controlled television eyes. Their hands have fingers which are equipped with pressure-sensitive pads to give them a sense of touch. And there are hundreds of machines that do routine (and even not so routine) chemical analyses, and that may therefore be said to have senses of taste. Machine production of fairly high-quality humanlike speech has been achieved, principally at MIT and at the Bell Telephone Laboratories. The U.S. Department of Defense and the National Science Foundation are currently supporting considerable efforts toward the realization of machines that can

understand human speech. Clearly, Simon's and Newell's ambition is taken seriously both by powerful U.S. government agencies and by a significant sector of the scientific community.

He also draws attention to a statement by Herbert Simon in his book *The Sciences of the Artificial* (1969):

An ant, viewed as a behaving system, is quite simple. The apparent complexity of its behaviour over time is largely a reflection of the complexity of the environment in which it finds itself... the truth or falsity of [this] hypothesis should be independent of whether ants, viewed more microscopically, are simple or complex systems. At the level of cells or molecules, ants are demonstrably complex; but these microscopic details of the inner environment may be largely irrelevant to the ant's behaviour in relation to the outer environment. That is why an automaton, though completely different at the microscopic level, might nevertheless simulate the ant's gross behaviour...

I should like to explore this hypothesis, but with the word 'man' substituted for 'ant'.

A man, viewed as a behaving system, is quite simple. The apparent complexity of his behaviour over time is largely a reflection of the complexity of the environment in which he finds himself... I myself believe that the hypothesis holds even for the whole man.[17]

Weizenbaum notes that 'With a single stroke of the pen ... the presumed irrelevancy of the microscopic details of the ant's inner environment to its behaviour has been elevated to the irrelevancy of the whole man's inner environment to his behaviour'.

ROBOTS AND SMART MACHINES

The concept of creating an artificial man or automaton capable of responding to orders dates back to the Jewish legend of the 'golem' in sixteenth-century Europe. In the eighteenth century Vaucanson – as described earlier in this chapter – and the father and son Jaquet-Droz made skilful automata in human form, but with a limited range of functions.[18]

The word 'robot' was coined by the Czech novelist and playwright Karel Čapek (1890–1938) in 1917 and popularized in his play *R.U.R.* (Rossum's Universal Robots), published in 1921. This described the impact of competition from robotic technology on human capacity, and its theme was taken up in Fritz Lang's film *Metropolis* (1926) and Charles Chaplin's *Modern Times* (1936). Robots have become a familiar staple in toys, comics, and films such as *2001* and *Star Wars*. However, modern industrial robots are not humanoid, do not converse, and are confined to carrying out highly specific tasks. Norbert Wiener drew attention to the paradox that the two qualities demanded of a slave – intelligence and subservience – are not compatible when taken to extremes. He wrote:

If our sole orders to the factory are for an increase in production without regard to the problems of unemployment and of the redistribution of human labour, there is no self working principle of laissez-faire which will make these orders rebound to our benefit and even prevent them contributing to our self destruction. The responsibilities of automation are new, profound and difficult.[19]

The most sophisticated robots so far have been used in space probes, such as NASA's Viking 2 which landed on Mars in 1978. However, the commonest are the industrial models such as Unimate and Puma. In the 1960s, typical assembly-line robots cost $US25 000, an equivalent of $4.20 per hour for a working lifetime of about eight years – slightly more expensive than a human worker. By 1980 a typical robot cost $40 000, but operating costs have only risen to $4.80 per hour while those of human labour have risen to $8 – $12 per hour. In the United States, it is estimated that the average cost of robots will fall to the equivalent of $10 000 by 1990.

Estimated numbers of industrial robots in operation (1979) are as follows:

Japan	10 000
US	3 000
West Germany	850
Sweden	600
Italy	500
Poland	360
France	200
Norway	200
Britain	185
Finland	130
USSR	25

As Japan cuts into the US car market, GM, Ford and Chrysler (if it survives) will be forced to robotize their factories. So will Australia, in the face of industrial competition from South-East Asia, unless tariffs and quotas are maintained. Australia has about one hundred industrial robots, largely because management is apprehensive about the social and political effects of job displacement but also because existing levels of production do not warrant expensive and wholesale conversion to new methods of production. On visiting factories in Australia I have been told repeatedly: 'If we needed to double or treble our output we would robotize.' Ominously, computerization enables robots to achieve parthenogenesis – they are able to reproduce themselves (asexually) and introduce modifications and improvements to each succeeding generation.

Office work will be increasingly dominated by 'smart machines' – computers which take, handle and process data without reference to humans. Many ostensibly labour-saving devices in the office have led to vastly increased output (and more incidental work) rather than work reduction – e.g. dry copiers (Xerox, Nashua and others) have not displaced secretaries – and it seems likely that word processors will have the same effect. CITCA put an unnecessarily heavy emphasis on word processors, and commissioned two special reports on this particular form of technology. It concluded (correctly) that word processors are not major factors in job displacement, but failed to examine the prospect of a paperless office with direct 'intelligent' contact from machine to machine – which would thus eliminate even the need for word processors (or their operators, manufacturers or distributors).

6 Unemployment, Inflation, Demand and Productivity

Now is the time to recognise that the real factors of production are energy, matter and knowledge, and that the output is human beings.

Hazel Henderson

Australia's national consciousness

The 1980s opened with Australia holding to a socio-economic policy line which could be described as a revolution of the radical right. The Fraser Liberal-NCP coalition is a beneficiary of the post-industrial revolution: the coalition and its allies – including multi-national corporations, technocracy, heavy industry, public service and controllers of the mass media – are, although some of them do not know it, a revolutionary class which has effected a major shift in Australia's politics. Its economic thinking is Friedmanite, dedicated to the supposition that increased market demand will eliminate unemployment as long as inflation can be contained.

Malcolm Fraser heads Australia's most conservative government since S. M. Bruce was Prime Minister in the 1920s. With Gough Whitlam's defeats in 1975 and 1977, the electorate moved to the right – taking the ALP with it. There are inbuilt institutional restraints against political or social reform – the Constitution, the Senate, the states and the media (which often trivializes or ignores serious political issues, except where personal confrontation is involved), and the ALP has been forced into a conservative, defensive role.

The Liberal Party is aggressive in its pursuit of foreign investment, and uses state power not as a national counterweight but in support of the multi-national corporations. It has adopted technological determinism with enthusiasm (see Chapter 10). On the technology issue, Australia is facing the worst of both worlds. An unresisted drive for the adoption of foreign-owned technological innovation is accompanied by a failure to grasp its economic or social implications, a dismal record in research and development, feeble management strategies and a prevailing lack of intellectual vitality in government or public service. The best to be said for a philosophy of 'muddling through' is that it might delay large-scale technological unemployment.

The ALP is still searching for a future role: Fraserism has taken the initiative and Labor has become a party of resistance to change without a dynamic of its own. The ALP has to develop policies which anticipate and will encourage massive social change to ensure that new technology is used to spread economic benefits more equitably. Neither major party, nor the Australian community, has evolved new policies which recognize the significance of possible energy crises and of structural trade dilemmas, especially with Asia. Australia has still to grasp that chronic unemployment could lead to the permanent alienation of the young, and that this would have pathological effects which would, in time, affect every family.

CLASS AND INEQUALITY

Many Australians regard themselves as members of a classless society, but when pressed by opinion pollsters to identify their social ranking, more see themselves as belonging to the middle class than do British or Americans, as illustrated in the following table:

Subjective class: Australia, Great Britain, US

Subjective class*	Australia(%)				Britain(%)		US(%)
	1961	1967	1972	1976	1963	1964	1970
Middle	62	54	62	68	32	41	46
Working	38	46	38	32	67	59	54

Source: David A. Kemp, *Society and Electoral Behaviour in Australia* (University of Queensland Press, 1978), p. 14.

* Professor Kemp notes: 'Totals have been percentaged to exclude respondents who did not place themselves in a social class. "Middle class" includes "upper class".'

On any *objective* analysis of wealth and income levels, the working-class section of the Australian community would be far greater than this table indicates. There may be more geese than swans in any particular area, but if half the geese regard themselves as swans then the swans will dominate.

Dennis Altman suggested that the myth of a 'classless' Australia arose from the high degree of homogeneity in social behaviour: 'Class differences are not immediately apparent from speech, dress or behaviour patterns . . . Constant reiteration of the theme "we are all middle-class now" serves . . . to disguise the realities of power relationships.'[1] A research report published by the Australian Commission of

Distribution of wealth and income in Australia, 1966–68

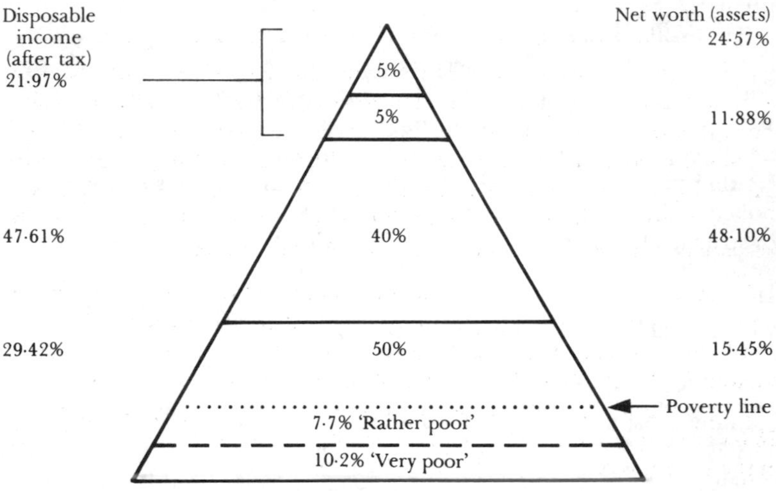

Inquiry into Poverty, chaired by Professor Ronald Henderson, provides a cautious analysis of wealth and income levels in Australia. In the diagram above, the top 5 per cent in the socio-economic pyramid may be described as 'upper-class', the next 5 per cent as 'upper middle-class', the 40 per cent below as 'middle-class', and the remaining 50 per cent as 'working-class'.[2] The designations 'very poor' (below the poverty line) and 'rather poor' (less than 20 per cent above the poverty line) were assessed by the Poverty Commission as at August 1973. There was probably little change in the relative size of wealth and income strata between 1966–68 and 1973. Between 1976 and 1980, the gap between rich and poor almost certainly widened because of galloping unemployment: the Australian Bureau of Statistics (ABS) found that 1·7 million income units were below the poverty line in 1979 compared to 1 million in 1973.

On the Poverty Commission analysis, while the middle-class share of income seems comparatively equitable, the discrepancies between the highest and lowest economic groups have disturbing social implications for the future. Phil Raskall has argued that the Poverty Commission understated the extremities of wealth and poverty, that the top 1 per cent own 22 per cent of the nation's wealth, and the top 5 per cent own 45.5 per cent.[3]

Employment shifts

It was the Whitlam government's misfortune to be in office (1972–75) when the structural effects of post-industrial change bit into employment, compounding existing income disparities – and doubly so that it had no historic or analytic grasp of what was happening (or why). Labor also came to feel that Keynesian economics was of limited and diminishing relevance, lost the faith, and helped to create an ideological vacuum which was filled by 'free market' philosophies.

In addition to the technological bite, there were four factors which helped to change the composition of the labour force:

1 A higher proportion of Australia's population than ever before was aged between fifteen and sixty-five, and offering for work – the result of unusually high birth rates in the 1950s. (These contracted steadily in the 1960s and more sharply in the 1970s.)

2 Increasing participation of women in paid work, rising from 26.6 per cent in 1966 to 42 per cent in 1980. (Marriage bars were lifted in the Commonwealth public service in 1967.)

3 The wage explosion of 1974 (including equal pay for women) which encouraged higher participation rates.[4]

4 The levelling off of educational participation rates in the 1970s. They fell in some areas, increasing the competition for jobs and leading to self-limitation for future employment.

Unemployment

In Britain at the end of World War II, John Maynard Keynes and William Beveridge defined 'full employment' as the equivalent of 3 per cent unemployment or less. To have 100 per cent employment would be 'over full', implying a rigidity in the labour market whereby people could not move in or out of particular jobs. Milton Friedman later asserted that in capitalist economies there was a 'natural rate' of unemployment of between 3 and 4 per cent, mostly workers in transition.

Ben Chifley's 1945 White Paper *Full Employment in Australia* set a goal which was maintained for thirty years. ABS figures indicate that between 1946 and 1974 the average unemployment rate was 1·2 per cent, and it became a political axiom that no government could survive more than 2 per cent out of work. In 1961, the Menzies government applied a credit squeeze to reduce inflation and unemployment rose briefly to 2·5 per cent: it survived the elections with its majority of thirty-two slashed to one seat (after electing a Speaker). From 1970, the unemployment rates as measured by the

Commonwealth Employment Service (CES) and the ABS for May each year were as follows:[5]

	CES (%)	*ABS* (%)
1970	1·0	1·3
1971	1·4	1·5
1972	1·9	2·0
1973	1·5	1·8
1974	2·1	1·6
1975	4·2	4·7
1976	4·3	4·4
1977	5·0	5·5
1978	6·1	6·2
1979	6·6	6·2
1980	6·5	6·2

Unemployment is an arbitrary and subjective concept, difficult to define and hard to measure accurately: 'No exact measure of "unemployment" exists. It can almost be said that unemployment is what it is defined to be.'[6] The 1911 census found that 16 per cent of boys aged between ten and fourteen were in the labour force: now the figure would be negligible, but we do not regard this as adding to our total unemployment. Similarly, the early censuses also included people of seventy or seventy-five in the labour force: now we do not. What of wives or mothers who lose husbands and whose children leave home: are they unemployed? What of those who are forced to leave work but wish to continue working after the age of sixty or sixty-five: are they counted? The rapid growth of part-time work adds a further complication. Between May 1964 and May 1972, full-time employment increased by an average of 2·2 per cent per annum and part-time by 4·5 per cent. Between May 1972 and May 1978, full-time employment increased by 0·6 per cent and part-time by 9·3 per cent.[7] Much part-time work is willingly chosen: much is not. 'Since August 1979 almost 16 per cent of all workers were employed part-time.'[8] In addition, the ABS estimated that about 3 per cent of the labour force (about 190 000 people) held second jobs and that this figure is relatively stable.[9] It is, however probably understated: many of those involved in moonlighting wish to avoid taxation and are thus discouraged from recording their additional employment, and it is possible that their number is approximately equal to that of the registered unemployed.

However, the published figures grossly understate the extent of Australia's work drought. For example, in May 1979, of 396 600 unemployed in the ABS survey, 28 per cent were not registered with the CES. In addition, CES records do not take account of under- or mis-employment. 'Discouraged job seekers' are a large and growing category, defined by ABS as persons who wanted a job but were not actively looking for work because they believed they would not be able to find a job for any of the following reasons: they were considered by employers to be too young or too old; they had language or racial difficulties; they lacked the necessary training, skills or experience; or there were no jobs in their locality or line of work.[10]

Falling CES registration is sometimes taken as evidence that the job market is improving. This is not necessarily the case: it may indicate a heavy fall in morale, a withdrawal from the labour market, and a sense that looking for work is futile. The Treasury paper *Job Markets* points to the extraordinary variation in the annual rates of increase in the labour force:

June 1961–June 1966	Average annual increase:	126 000
May 1966–May 1971	Average annual increase:	139 000
May 1974–May 1978	Average annual increase:	88 000
Nov. 1977–Nov. 1978	Actual increase:	25 000

Figures supplied by the Treasurer, John Howard, in answer to a question on notice (5 June 1981) indicate extreme variability in labour-force increases for the years 1966–80: the lowest number was 10 600 (year ending August 1978) and the highest 223 700 (to August 1980). This suggests that 'registered unemployment' figures are almost meaningless unless we know the true size of the potential labour force.

Dr Duncan Ironmonger, of Melbourne University's Institute of Applied Economic and Social Research, attempted to estimate a 'true unemployment' figure by examining recent trends in increased labour-force participation in Australia, and taking account of the demographic bulge in the proportion of people of labour-force age. He concluded that for 1980 the actual employed – both surveyed and hidden – amounted to 971 000, or 13·77 per cent of the potential labour supply.

A survey by the Commonwealth Parliamentary Library concluded that actual unemployment and under-employment in May 1978 was about 12·5 per cent of the potential labour force (840 000 people). Keith Windschuttle estimated that the total level of unemployment, under-employment, disguised and concealed unemployment in Australia was about 18 per cent of the potential labour force – i.e. 1·25 million people wanted to work full-time and could not.[11] In the 1979 Estimates debates for the Department of Employment and Youth Affairs, Ian Viner indicated that he could not reject or confirm larger

Employment, unemployment and potential labour supply
(Averages for calendar years, '000s)

	Employed surveyed*	Unemployed			Potential labour supply*
		Surveyed*	Hidden	Total	
Actual–					
1971	5 517	107	..	107	5 624
1972.........	5 602	150	..	150	5 752
1973.........	5 765	136	..	136	5 901
1974.........	5 891	162	..	162	6 053
1975.........	5 867	303	39	342	6 209
1976.........	5 946	298	125	423	6 369
1977.........	6 000	358	175	533	6 533
1978.........	5 974	409	319	728	6 702
1979.........	6 064	396	415	811	6 875
Forecast–					
1980.........	6 080	462	509	971	7 051

* Potential labour supply equals surveyed labour force up to 1974, thereafter assumed to grow at 2·6 per cent a year (2·1 per cent for growth of population aged 15 and over, and 0·5 per cent for increased participation in the labour force.

Source: ABS, *The Labour Force, Australia* (Catalogue no. 6203·0 and no. 6202·0)

estimates of total unemployment.[12] The Australian economists Bob Gregory and Ron Duncan pointed to the paradox that in previous recessions labour participation rates have tended to fall quickly, which increases 'hidden unemployment' but decreases registered unemployment. However, in the 1974–75 recession participation rates *increased*, leading to a dramatic rise in registered unemployment.[13] Aggregated Australian unemployment figures are serious enough, but they must be disaggregated to demonstrate the segmented nature of changes in employment.

There has been massive failure by politicians, journalists, public servants and professional economists to look at, let alone recognize, structural changes in employment. The following points have been insufficiently analysed:

1 *The class nature of unemployment.* Unemployment is four or five times more serious in working-class than in middle-class areas – and the psychological and economic effects are even more damaging, since self-definition and feelings of personal security in the working class are more strongly determined by employment.

2 *The regional nature of unemployment.* CES figures indicate that

unemployment is relatively low in inner-city areas or in the suburbs surrounding the inner city, compared to the very high levels in outer suburbs and in country districts – particularly where employment is over-specialized (e.g. factory employment and agriculture) and economies of scale operate against job creation.

3 *The ethnic bias in unemployment.* Unemployment rates are highest in areas with heavy migrant populations. The lack of language skills is especially serious.

4 *The youth bias in unemployment.* Young people, especially those who have been poorly educated or specifically trained for jobs where demand is falling, are a glut on the market. The CES reported that in Victoria in July 1980, the ratio of registered unemployed to unfilled vacancies was 41·6 to one for juniors and 17·6 to one for adults.

5 *The sex bias in unemployment.* Employment prospects for females in city areas are slightly better than the male adult rate, but in country areas job prospects are poor, especially for juniors.

In St Albans, an outer Melbourne suburb in the electorate of Lalor (for profile, see Chapter 3) with large numbers of young people, a heavy migrant concentration, low educational participation rate, and over-specialized, highly efficient industries – there were 1776 registered unemployed in August 1980 and only thirteen vacancies – a ratio of 136·6 to one.

There are two apparently contradictory elements in employment trends in Australia:

1 Shortage of skilled workers, including tradesmen and professionals such as computer programmers.

2 Surplus of prospective job applicants with minimal skills or no skills at all, for whom neither jobs nor incomes can readily be found.

There are pages of classified advertisements in daily newspapers asking for skilled personnel. This is often taken as proof that there is no serious unemployment problem in Australia, and it is common for employers to say 'I have been advertising for skilled workers for months and I cannot fill the jobs.' On the other hand, when white-collar jobs are advertised it is common to receive scores and sometimes hundreds of applicants, many of them educationally over-qualified. Declining opportunities for satisfying work for the semi-skilled and unskilled will become Australia's greatest social problem in the 1980s.

Most technologically advanced nations experienced high unemployment rates in the 1970s, as illustrated in the table on p. 131. In

Exit Full Employment, Dr Barry Hughes wrote: 'Sweden, Japan and Austria have been notably successful in resisting the international recession . . . Full employment is alive and well, but it has been forced to live in Austria, Japan and Sweden.'[14] There are particular reasons for low unemployment rates in these three countries, and also in West Germany:

Austria. The International Labour Organization (ILO) reports that Austria has 15 per cent of its labour force in agriculture, which reduces competition for available places in cities and towns. Rural under-employment is concealed on the farm. The Austrian working week in manufacturing industry is only 34·4 hours, and there is a high employment rate of 30·4 per cent. The Austrians have been into *de facto* work sharing for decades. The socialist government of Chancellor Bruno Kreisky has amiable relations with the trade unions and has evolved a successful incomes policy. Austria also has by far the largest number of tourist arrivals of any major country (1544 tourists for each 1000 inhabitants, compared to Australia's thirty-nine per 1000). However, it must be conceded that educational participation rates are relatively low.

Japan. Contrary to the popular stereotype, 20 per cent of Japan's workers are farmers. The normal retirement age for urban workers is fifty-five or fifty-seven (lower for women), and people seeking jobs above that age are not recorded statistically (see also Chapter 9). As part of the tradition of lifetime employment, many under-employed or technologically redundant employees are carried by firms as 'window-gazers' – clocking on and off with the other workers, but performing no duties. If Japan used Australian criteria to measure unemployment, its figures would be in the region of 7–9 per cent.

Sweden. Sweden has a comparatively large industrial labour force (26·5 per cent), a very short working week (30·2 hours), a craft tradition of relatively low-volume high-quality production based on indigenous technological development, and an unusually large proportion of people employed in welfare services.[15] In addition, the practice of industrial democracy and consultation between governments, employers and unions is firmly established.

West Germany. West Germany has been able to export unemployment: between 1973 and 1977, 650 000 'guest workers' were repatriated, largely to Yugoslavia and Turkey. In the same period, about 300 000 workers took advantage of early retirement schemes. In the 1973–75 recession a maximum of 773 000 workers were supported by the

Kurzarbeit scheme, by which two-thirds of lost wages were made up from government and EEC funds. In addition, some industries have adopted a shorter working week. A total of 1·7 million jobs were lost 1973–77, but the labour force fell by 1 million.[16]

Unemployment as a percentage of civilian labour force

OECD countries	1974	1975	1976	1977	1978	Jan.† 1979	Absolute 1978 number ('000)
Belgium	3·2	5·3	6·8	7·8	8·4	8·5	282
Denmark	2·0	4·6	4·7	5·8	6·6	6·7	191
France	2·3	3·9	4·3	4·9	5·3	5·8	1167
Germany	2·2	4·2	4·1	4·0	3·9	3·7	993
Ireland	6·3	8·7	9·8	9·7	8·9	8·3	102
Italy	4·9	5·3	5·6	6·4	7·0	7·1	1571
Luxembourg	0·0	0·2	0·3	0·6	0·8	0·8	1
Netherlands	2·9	4·1	4·4	4·3	4·3	4·3	206
UK	2·4	3·8	5·3	5·7	5·7	5·4	1376
Canada	5·5	6·9	7·1	8·1	8·4	8·1	922
USA	5·6	8·5	7·7	7·1	6·0	5·8	6047
Japan	1·4	1·9	2·0	2·1	2·3	2·1	1240
Australia	2·1	4·4	4·6	5·6	6·3	7·0*	402
New Zealand	0·1	0·3	0·4	0·6	1·9	NA	22
Austria	1·4	1·8	1·9	1·8	2·1	3·5*	59
Finland	1·7	2·2	4·0	6·1	7·5	7·6	169
Norway	0·6	1·1	1·1	0·9	1·1	NA	20
Spain	3·4	4·9	5·6	6·8	9·1	NA	1095
Sweden	2·0	1·6	1·6	1·8	2·3	2·8*	94
Turkey	7·0	7·0	8·3	NA	NA	NA	NA
Switzerland	0·0	0·0	0·4	0·9	0·5	NA	10.5

* Seasonally adjusted.
† Not adjusted.

Source: (For the EEC countries) EEC, *Recent Economic Trends* (no.3, March 1979); (for the other OECD countries) calculated from OECD, *Labour Force Statistics, 1965–76* (Paris 1978) and the quarterly Supplement (February 1979/1), as well as *Main Economic Indicators* (1978/2 and 1979/4).

INFLATION WITH UNEMPLOYMENT

In the era of full employment (1945–74), conventional wisdom asserted that there was a 'trade-off' between employment and inflation. At its simplest this meant that, in times of prosperity, higher spending + higher employment = faster inflation, while lower spending + lower employment = reduced inflation. This relationship – first described in 1926 by Irving Fisher – was elaborated by the economist A. W. H. Phillips, who devised 'the Phillips curve'.[17] He argued that there was an inverse relationship between the rate of price (including wage) rises or falls and unemployment levels.

In Australia until 1974, unemployment ran at 1 per cent or less, while inflation generally ranged between 2 and 3 per cent. In 1951, as a result of enormous increases in world commodity prices triggered off by the Korean War, Australia's inflation rate surged to 25·2 per cent with unemployment at only 0·3 per cent. It fell back to 3·5 per cent in 1952, with an average of 1·2 per cent unemployment (rising to 2 per cent for one quarter). Between 1961 and 1970, Australia had an inflation rate average of 2·5 per cent per annum, compared with 3·4 per cent for the OECD, 4·1 per cent in the UK and 2·8 per cent in the US. This conformed to the classic Phillips curve, where wage inflation (W) is measured on the vertical axis, and the unemployment rate (U) on the horizontal:

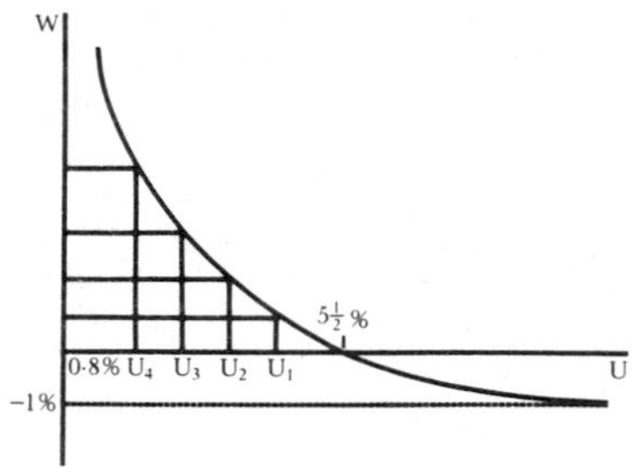

The curve is non-linear, successive reductions in the employment rate result in ever higher increases in the rate of wage inflation. For example, a reduction in the unemployment rate from U_3 to U_4 will involve a much larger increase in the rate of wage inflation than would an *equal* reduction in the unemployment rate from U_1 to U_2 . . . As U approached 0.8 per cent, wage inflation would approach infinity . . . Once the curve passes below the horizontal axis it becomes perceptibly flatter. This has often been regarded as an empirical vindication of Keynes' view that even at high rates of unemployment, money wages would not fall to any appreciable extent. The *minimum* rate of wage inflation predicted by the Phillips curve was of the order of minus one per cent.[18]

The Phillips relationship has been true in most recent *expansions*: i.e. when prices have risen, unemployment has fallen. However it does not appear to be true in *contractions*: in recent recessions, unemployment and prices have *both* increased. Of course, the Phillips principle does not claim that prices fall in recessions, only that the *rate* of price increase would be lower as unemployment rose. By the 1970s, governments had faced two radically different phenomena, slumpflation and stagflation – similar but not identical combinations of high unemployment, high inflation and low production growth rates. *Slumpflation* is marked by a high rate of unemployment increase, with a simultaneous rapid rate of price increase. *Stagflation*, common from the 1970s on, involves a more or less constant rate of inflation at each change of unemployment. It is possible that the inflationary nature of economic life in future will settle down to a steady rate of price increase accompanying a moderate but consistently increasing rate of unemployment. Keynesian theory proposed that a growing (or decreasing) employment rate would induce an increasing (or decreasing) rate of prices (including wages). From 1945 to the 1960s that was the experience, particularly in the upward direction, but this has not occurred since.

In the 1960s, the US economists Milton Friedman and Edmund Phelps independently devised a theory of 'inflationary expectations'. When people living in a full-employment economy accept inflation as inevitable, they incorporate this in their individual or collective decision-making. For example, if wage negotiations have to allow for inflation, then unions incorporate inflationary expectations into demands and employers then take a wage settlement as a green light for further price increases. People invest in homes or cars which appear to be beyond their reach, because they expect to pay for them out of inflation-generated wage rises. This becomes a self-fulfilling prophecy. The Phillips curve has been redefined as the 'Friedman-Phelps expectations-augmented Phillips curve'. As J. A. Trevithick writes:

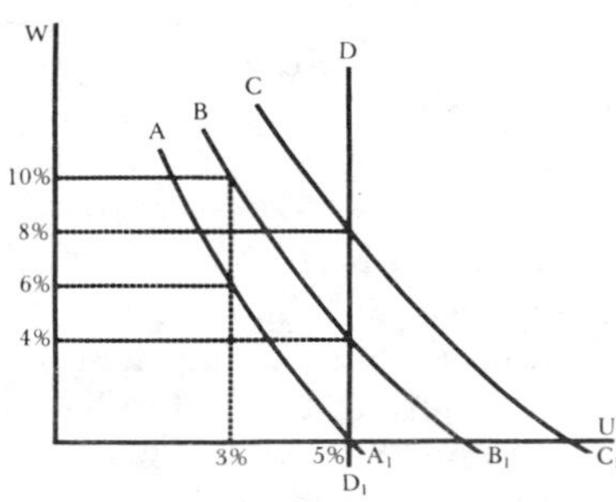

A given Phillips curve . . . cannot be regarded as a stable relation whose existence into the foreseeable future can be guaranteed. There will be an infinite series of short-term Phillips curves, each one corresponding to a different expected rate of inflation . . . The AA_1 curve corresponds to a zero expected rate of inflation, the BB_1 curve to a 4 percent rate: and the CC_1 curve to an 8 per cent rate . . . In the long run the only choice which the authorities face is which point along the vertical line DD_1 they are eventually prepared to settle upon.[19]

In the 1970s, inflation rose sharply in most technologically advanced countries, from an OECD average of 3·4 per cent for the period 1961–70 to a high point of 13·4 per cent in 1974. Japan's inflation peaked at 24·5 per cent in 1974 and the UK at 24·2 per cent in 1975, while Australia was at 15·1 per cent in both 1974 and 1975. Conventional economic wisdom links the growth of inflation in the 1970s, especially in Australia, with wage rises and deficit budgeting. A comparison of inflation – as measured by consumer price index (CPI) figures – increase in average weekly earnings, and the size of domestic budget surpluses (or deficits) for the 1970s suggests that the links are apparent in some years, but not always (even allowing a time lag of twelve to fifteen months for flow-on to occur):

Year	*Inflation(%)*	*Wage rises(%)*	*Budget surplus (+) or deficit (–)*	
1970	4·9	8·4	$201m+	(1969–70)
1971	7·2	10·3	$519m+	(1970–71)
1972	5·9	11·0	$405m+	(1971–72)
1973	9·5	8·9	$215m+	(1972–73)
1974	15·1	16·2	$211m+	(1973–74)
1975	15·1*	25·4	$1949m–	(1974–75)
1976	13·5	14·4	$2873m–	(1975–76)
1977	12·3	12·4	$1865m–	(1976–77)
1978	8·2	9·9	$2361m–	(1977–78)
1979	9·1	7·7	$2258m–	(1978–79)
1980	10·8	9·5	$567m–	(1979–80)

* Inflation was at its highest in January-March of 1975, when it reached 17·6 per cent (falling to 13–14 per cent within twelve months).

The inflationary increases of the 1970s, for whatever reasons, caused a political backlash to the right in Australia, the United Kingdom, the United States and Israel; the OPEC oil price rises were another major factor. In Australia, the trade unions and their wage demands were denounced as the prime internal cause of inflation, a line which was pushed with enthusiasm by the Liberal Party, the NCP, the mass media, the corporate sector, and senior officers of the public service (of whom John Stone, Secretary of the Treasury, was most prominent). In addition, the Australian Treasury adopted the monetarist argument of Milton Friedman and Friedrich von Hayek that inflation and thereby wage inflation could only be controlled by governments if they cut back on money supply.[20] Keynesians argue the more subtle point, however, that when wage pressure is created, money supply increases to meet it—it is not the other way around. Despite the statistical association of inflation and wage increases, there is no proof that expanding the amount of money available permits wage increases: rather, the supply of money is itself a function of the price level, and the price level that of the wage level.

In the 1975–76 Labor budget, the then Treasurer Bill Hayden adopted a 'Beat inflation first' strategy which had a short-term success in reducing cost rises. The incoming Fraser government imposed 'incomes policies' to prevent workers from securing gains in real wages, and rises were indexed to CPI figures. These policies were circumvented by 'sweetheart' arrangements between unions and employers, and trade-union hostility to government policies disrupted the economy. Cutbacks in government employment did not lead to the hoped-for generation of new work in the private sector. High unemployment contributed to high taxation for unemployed relief, and discouraged investment through falling consumption. Reductions in medical and hospital benefits through Medibank simply led to increased private medical expenditure which had a significant inflationary impact on the CPI. So did the government's decision to pass on OPEC's increased fuel prices to Australian consumers – purported to be a means of curbing demand (largely unsuccessful), but in fact aimed at raising revenue.

In attempting to explain the relationship between inflation and unemployment, insufficient attention is being paid by governments to the major structural changes in jobs which constituted the post-industrial revolution of the 1970s. Robert Bacon and Walter Eltis at least noted the principle – recognizing that Britain's inflation was compounded by the relative decline in market-sector employment and the growth of the non-market sector – but they failed to grasp that Britain's problems would be made worse if the historical trend were reversed.[21] British Leyland had too many workers, not too few: it would have become bankrupt if it had employed more; similarly British Rail and British Post would be weakened by more employees. The dilemma is shown: more public-sector spending leads to inflation; more private-sector employment attempted leads to unemployment.

The sheer complexity of transactions in a technological machine-dependent society has been a major, although largely unrecognized, element in inflation. This complexity leads to an inflationary cycle:

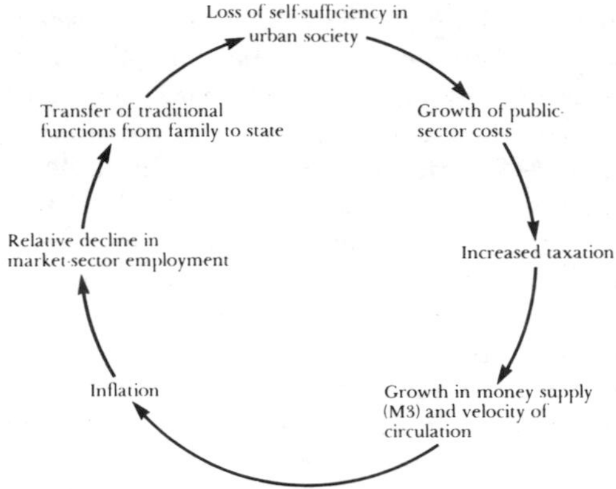

In 1974 the Canadian government commissioned GAMMA, a research group sponsored by Montreal and McGill Universities, to study the impact of uncontrolled, undirected industrial growth in Canada and to propose alternatives. In 1976 GAMMA published a four-volume report, *The Selective Conserver Society*, comprising a one-volume synthesis and fifteen supporting technical papers. It examined five different scenarios for the future, the first two being models of the consumer (or 'cornucopian') society:

1 The status-quo scenario (CS_0) – i.e. doing more with more: business as usual.

2 The squander-society scenario (CS_{-1}) – i.e. doing less with more: waste and extravagant consumption.

They then offered three models of the 'conserver' society:

3 The growth-with-efficiency scenario (CS_1) – i.e. doing more with less.

4 The high-stable-state scenario (CS_2) – i.e. doing the same with less.

5 The post-industrial-conserver scenario (CS_3) – i.e. doing less with less and doing something else.

The report concluded that existing Western economic models ('cornucopian' societies) are inherently inflationary: excessive use of resources to meet desired ends (e.g. emphasis on moving vehicles rather than people) inevitably increases costs and overheads, and exacerbates the tendency towards loss of self-sufficiency. The underlying structural cohesion of the economic system is at risk – quite apart from additional shocks imposed by the technological revolution.

The report concluded that a stable-state economy ('conserver' society) in which the value of total inputs and outputs was roughly equivalent would have anti-inflationary tendencies. A transition from a post-industrial to a post-service society could lead to a more intimate relationship between the producer and consumer of goods and services, a greater degree of self-sufficiency and lower inflation.

GROWTH, DEMAND, AND EMPLOYMENT

A fundamental tenet of Keynesian and neo-Keynesian economics is that employment levels are determined by: (a) demand for goods and services, and (b) rates of growth. Macro-economic adjustments, 'fine tuning', or 'priming the pump' can restore employment to the desired state.

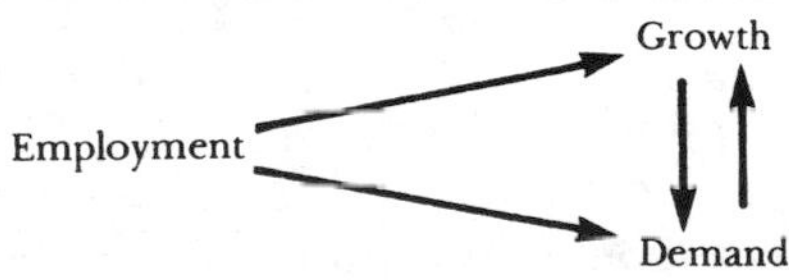

Friedman's dictum that a series of short, sharp shocks would lead to a revival of private investment in industry, end feather-bedding and restore employment to a 'natural' level was applied with vigour by Margaret Thatcher's Conservative government after 1979: Sir Geoffrey Howe, Chancellor of the Exchequer, and Sir Keith Joseph, Secretary of State for Industry, were convinced monetarists. By 1981, however, Britain had an inflation rate of 22 per cent and unemployment stood at 10·9 per cent, the highest figure since the 1930s.[22]

Keynesians argue that since 1974 there has been an international slump marked by rising levels of unemployment, especially in manufacturing. They presume that if the slump ended there would be a restoration of full employment and a revival of manufacturing *as an employer*. There is a circularity in this argument: the main evidence for S is U, and the reason for U is S. In Australia, at any rate, there continue to be very high and rising levels of demand for most goods and services – which suggests that changes in *modes* of production make the *numbers* of people employed of diminishing relevance. Further, the decline in Australian manufacturing began in 1965, not 1974.

Okun's Law, propounded in 1958 by the American economist Arthur M. Okun, states that the unemployment rate moves by one-third as much as the gap between actual and potential real GNP.[23]

This law, which seemed completely relevant in the 1960s, has been revised in the 1970s to suggest that the responsiveness of unemployment to the gap is closer to 45 per cent than 33 per cent. The law has also been stated as follows:

> For the unemployment rate to remain constant, a level of output must grow at a rate equal to the growth rate of the labour force plus the growth rate of productivity per worker. Output growth that falls short of this sum will push unemployment upwards.[24]

Okun's Law and the Keynesian belief that demand *of itself* leads to increased employment are not true for high-volume capital-intensive industries (or the regions where they are located). The implication that growth in X (aluminium smelting) would necessarily lead to increased employment in Y (teaching), for example, must be challenged. In Great Britain the highest unemployment is in the industrial Midlands, the lowest in the service-based south-east. The Lalor electorate, with the highest 'value added' per worker for any part of Australia, should have the lowest unemployment if Okun's Law is correct – in fact, it has about the highest. Western Australia, whose economy is dependent on the exploitation of natural resources, has the highest unemployment rate of any state. US Department of Labor statistics confirm that between 1973 and 1979 the American economy created jobs at three times the rate of Japan – faster than any other major industrial country – yet the United States had an unusually low rate of growth while Japan's was relatively high. In addition, Okun's Law begs several questions: How is growth measured? What is productivity? How are seasonal variations and voluntary part-time work taken into account? Since 1975, there has been no apparent correlation between annual rates of growth in GDP, inflation and unemployment in the US, as the following figures indicate:

Year	*Growth rate* (%)	*Inflation* (%)	*Unemployment* (%)
1976	5	5	8
1977	6	7	7
1978	5	9	6
1979	1	13	6
1980	2	18	6

Labour/time-absorbing employment remains highly sensitive to demand fluctuations, especially in the provision of personal services: if the demand for haircuts or tooth-fillings doubles or halves, we can expect the number of hairdressers or dentists to rise or fall. Labour/time-saving employment, on the other hand, is now much less sensitive to such changes. It appears paradoxical that small-scale increases in demand can raise employment while large-scale demand can lower it.[25] For example, if a garage proprietor has two pumps and one employee and doubles the number of pumps, he may well double the number of employees; but if demand increases by 400 or 500 per cent so as to justify eight or ten pumps, it is probable that the proprietor will go into self-service with electronic recording of transactions and retain only two employees. I propose Jones' Third Law:

In the production of goods or market services on a massive scale, employment tends to be in inverse proportion to demand.

The following examples illustrate the point:

1 US agriculture is the most abundant in world history, but its share of the labour force has fallen dramatically – from 38 per cent in 1900 to 3 per cent in 1978.

2 Supermarkets may have the same labour force as three or four family-run mixed businesses, but the turnover might be more than ten times as great. The use of UPC is intended to replace personnel in shops by automatically totalling the cost of purchases and making appropriate adjustments to inventories.

3 Japan, with large production runs and high technology, produces an average of 94 cars per worker per annum – but Toyota, the largest producer, has a far lower labour content than Mazda. The Nissan factory has an assembly line two kilometres long, in which work is 97 per cent automated and 3200 workers produce 420 000 vehicles per annum (an average of 131·25 per worker). In the Australian motor industry, 360 000 units are manufactured by 80 000 production workers, an average of 4·5 per worker. (If Australia increased its production runs to compete on world markets the existing ratios could not be retained.)

4 Exxon leads GM as the world's greatest capitalist enterprise – but its labour force is only one-seventh of GM's.

5 Sales of petrol doubled in Australia between 1970 and 1977, but employment of garage-pump attendants fell by 40 per cent.

6 Mexican agriculture is doubling its output at a time when the agricultural labour force is being dispossessed.

7 Banking and insurance transactions are up, while employment prospects are down. A single computer in Massachusetts now does all the recording and retrieval work for all banks in that state. In the ANZ Bank (Victoria), 480 personnel retired in 1978 to be replaced by 200 new employees in 1979.

8 Telecom expects to double its throughput and reduce the cost

of calls in the next ten years with only a 10 per cent increase in its labour force. The Telecom dispute of August 1978 directed public attention to the need to plan for job displacement by technology.

9 The Victorian State Electricity Commission (SEC) doubled its output between 1963 and 1972, while the production labour force fell. (An overall increase of 4·4 per cent occurred mostly in sales and promotion.)

10 The explosives and petrochemical industry in Australia increased output by 58 per cent from 1968 to 1977, while employment steadily fell from 10 169 to 9915.

11 Due to containerization, throughput on the Australian waterfront increased by 600 per cent from 1956 to 1978 while the labour force fell by two-thirds.

12 An English factory using semi-automated methods to manufacture taps and dies employs 300 employees, while an automated Japanese firm employs less than ten for an equal output.

13 Nine thousand workers in Japan produce as much steel as 100 000 workers in Britain. How much more steel would Britain have to produce to justify a pro rata increase in her steel labour force?

14 The total volume of printed material in the US has risen 300 per cent since 1950, but numbers employed in printing trades have fallen. Between 1967 and 1972, output increased by 31 per cent and the labour force dropped by 32 per cent.

15 Average output per employee in the Australian hosiery industry increased from about 2000 dozen pairs in 1973 to about 3400 dozen pairs in 1977–78, and employment has fallen from 5662 to 3972.

16 Worldwide employment in the watch industry has fallen from 32 000 in 1970 to 18 000 in 1977 – but output has increased and prices have generally fallen.

17 The NCR standard mechanical cash register had 370 moving parts: its electronic equivalent, cheaper and more efficient, has only seventeen (less than 5 per cent) and demands far less labour to manufacture.

18 Philips, the world's largest electrical manufacturing company, estimates that even after allowing for a 3 per cent real increase in annual turnover, by 1990 it will be over-manned by 56 per cent.

19 Production of colour television sets in Japan increased from 8·4 million in 1972 to 10·5 million (an increase of 25 per cent) in 1976, while the numbers of employees making them fell from 47 886 to 25 677 (a fall of 46 per cent).

20 In the United States, most bookings for major airlines are now carried out in San Francisco or Denver: prospective travellers in New

York, Miami or Chicago pick up their telephones and dial 'toll-free' because it is far cheaper and more energy-efficient for airlines to pay transcontinental telephone charges than to set up 100 offices with appropriate staffing in 100 cities throughout the country.

The historic relationship between output and total inputs – including employment, raw materials and energy – appears to be breaking down in many industries. Historically, output tended to increase at a faster rate than employment but a reasonably close – even if slowly diverging – relationship remained. The micro-electronics revolution has overturned many of the basic assumptions underlying production, by breaking the nexus between the cost of labour and technology.

Changes in the relationship between input and output

Historic relationship

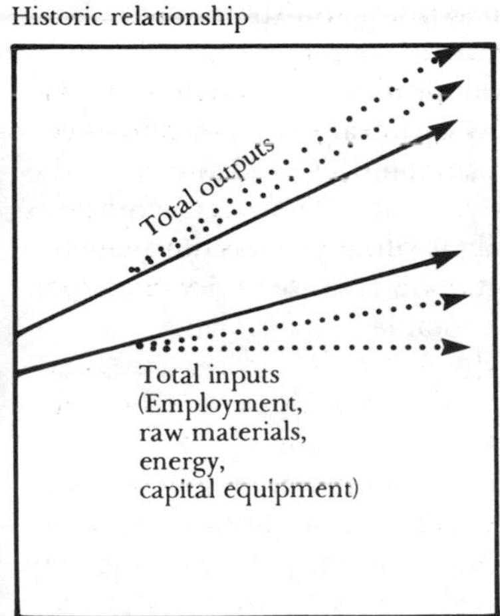

Impact of miniaturization

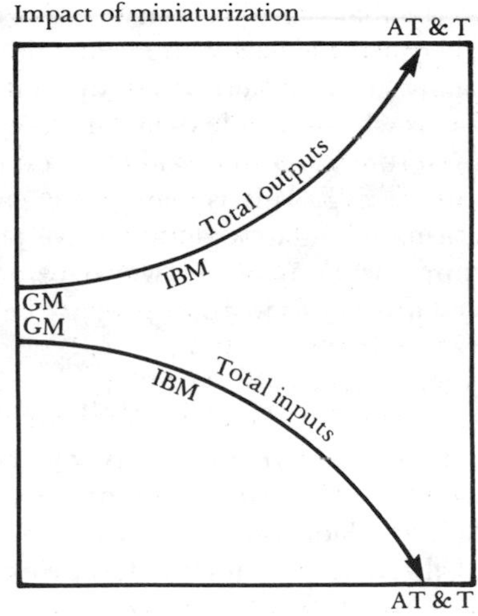

The input-output relationship of General Motors remains close to the historic mode. Its outputs are huge – but so are its inputs. In addition to its own large labour force, GM employs hundreds of thousands indirectly in building huge factories, presses, forges and furnaces, in providing steel, glass, paint, tyres, batteries, and in generating energy. Sales and promotion is not carried out by GM's own labour force (839 000 worldwide in 1978). By comparison, IBM's labour force is much smaller (see Chapter 5), so are its factories, and the volume of raw materials and energy used are modest. The largest section of IBM's labour force is employed in sales, promotion, research and producing software. AT & T (American Telephone and Telegraph) has even lower inputs compared to the value of its outputs in telecommunications.

Does greater output necessarily mean more jobs? Consider the following scenario:

Factory A has fifty employees on a 40-hour week, with no overtime and low-level technology, and produces 2000 garments each week.

Factory B has forty employees on a 40-hour week, with an average of ten hours per week in overtime, and produces 2000 garments each week.

Factory C has twenty employees on a 40-hour week, with high-level technology, and produces 4000 garments per week.

If the total demand for garments rises to 10 000 per week, we cannot assume that Factories A and B will employ more workers, especially if Factory C can make up the additional production at little extra cost. It is true that some jobs can be created in existing industry by cutting out overtime and filling the shortfall with new workers. However, since many householders have adjusted their standard of living to income levels which include overtime, this may be unsaleable to the workers. In Europe it is often said that 'job creation and work sharing' are the same thing. This is seen most clearly in Britain, where work that normally might be done by five people is often carried out by eight, simply working at a slower pace. But it *is* job creation – depending on low wage rates which would not be acceptable in Australia.

Job creation or reduced demand?

If Duncan Ironmonger is correct in his estimate that in 1980 there were 7 million Australians of labour-force age (and wanting work) and only 6 million jobs available, then to restore full employment a variety of alternative strategies must be explored:

1 Create 1 million more jobs.
2 Reduce by 1 million the number of people seeking work.
3 Achieve an appropriate mix of job creation and reduced work demand.

Among the 6 million in work, there are many who would do other things – study, return to domestic work, travel, pursue hobby and craft interests – but for income dependence and the fear of being regarded as useless and worthless by withdrawing from work. On the other hand, there may be nearly a million people who are desperate to obtain work, under almost any circumstances. The most humane way to handle labour-force problems is to assist those who want to get out of work to do so without trauma, and to provide income support, while encouraging those who want to get into work to do so.

In addition, the sheer productive capacity of our society and its dependence on increased levels of consumption will require that overall purchasing power is not allowed to fall in the long term. I propose Jones' Fourth Law:

The economic viability of a technologically advanced society depends on having an increasing number of small consumers despite a contracting number of large producers.

To assist citizens to carry out their patriotic duty to consume at high levels, it will be necessary to introduce some form of income support to enable the submerged one-sixth of the Australian community to raise consumption patterns towards the national average.

Appropriate strategies for encouraging untraumatized work transition and maintaining income spread are on the one hand to introduce a guaranteed income and/or national superannuation, or on the other hand to:

1 encourage more working-class and lower middle-class participation in education;
2 recognize the economic and social importance of domestic work;
3 promote employment based on leisure, tourism and craftwork.

Strongly dramatized conflicting opinions about job creation are pushed by both the right and the left. The *right-wing fantasy* asserts that if government expenditure, employment, taxation and inflation were all reduced a 'born-again' private sector would miraculously expand, with massive job creation providing unspecified new products and services within each national economy. The *left-wing fantasy* asserts that unlimited jobs can be created if there is only sufficient will and government money to do it. The chant often heard at unemployment demonstrations – 'What do we want? Jobs! When do we want them? Now!' – is not, as it may seem, the modern equivalent of a rain dance. In the short run, job creation is comparatively simple – extra people can always be employed to pick up litter, clean houses or run errands. The real problem is income creation. This income must become available from taxation, from money printed by the government, or from increased loan-raising from the public non-bank sector, and thereafter be derived from consumer spending. The new jobs must be wanted, useful, and fit into the national system without interfering too much with the prices and costs precariously in balance with our international competitiveness. It seems essential to point to Sweden, which has the world's highest living standard, one of the best employment records, *and* the highest percentage (53·5) of national income taken in taxation.

Let us pull something fundamental out of these opposing claims to the truth. Certainly to create jobs we must find the money, we must have ready the forms of new work to employ the unemployed in, and we must not cause any further inflationary pressure whatever. What can be done? One firm solution is for government to increase its expenditure – creating jobs – and simultaneously gather augmented tax to such an extent that the added spending is covered.[26] This is a tax policy from which members of parliament may naturally shrink despite its certainty: can it be set going in Australia? (The cry is for 'less government', but the practice we see around us in Western democracies is 'more government'.) Briefly, it would work as described in the following paragraph.

Taxation is raised to a degree that compensates for changes in government expenditure, to create jobs, right at the start. If you look closely at the circulatory effects of taxes and public spending, you must conclude that both national income and employment levels have increased. Yet the amount of money chasing consumption and investment goods has not changed one iota. *There is no added inflationary pressure because there is no extra money created.* These things happen because households receive exactly the same total income as before, but it is divided differently among them – taxpayers have less, new public servants have incomes. However, national production has increased; more government expenditure has been added to the same consumption and investment as before; economic activity has increased. The significance of this economic operation is that you do not have to unlock the gate on expenditure in order to ameliorate or cure unemployment. The poignant lesson is that governments must not reduce their total spending in a recession. The anti-spending policy statements of the Reagan, Thatcher, and Fraser regimes do not hold water and cannot persist in practice, if chronic unemployment is to be eliminated.

NEW WORK FORMS

As the labour content in high-volume production decreased in the 1970s, additional forms of work evolved. Most were complementary to, and not dependent on, new technological forms; some were based on the development of new products (e.g. male cosmetics, saunas, 'health foods', electronic games, Rubik's cubes, Jacuzzi baths and pool accessories); but even more work was generated by the creation of new services. As taxation was considered overburdensome and tax laws more complex, tax avoidance became a growth industry leading to the expansion of law firms and the recruitment of many tax-enforcement officers. Almost a million US inhabitants are employed in monitoring, cleaning up and eliminating various forms of pollution. Increasing delays and uncertainties in regular postal services in Britain have led to a dramatic increase in personal delivery and courier services. Increased income and leisure have contributed to the production of bad goods as well as good goods, and bad services as well as good services – there are the drug trade (and its detection) prostitution, massage parlours, gambling, heavy drinking as well as expanded use of hotels, motels, restaurants and fast-food chains. Some of the new employment required higher levels of skill than the jobs which were eliminated, serviced individual needs, and was labour-intensive. Unfortunately, the new work did not necessarily spring up in localities where old jobs were eliminated – housing and schooling anchor families to particular areas.

The American writer Lenny Kleinfeld has suggested some new job possibilities (not completely tongue-in-cheek), including computer psychiatrist, surrogate golfer, live audience member, sex-determiner, Kennedy aide and Federal croupier. But the most promising is that of scribe: 'Any day now, large segments of the increasingly illiterate population will require someone to read their mail for them and write out replies. College graduates will need someone to fill in their job applications.'[27]

Most economists have signally failed to recognize the significance of the *locality* in determining employment levels: its culture, ethnicity, education and environment are inextricably linked to what kinds of jobs are generated and where. Economics is no longer the science of 'what, how and for whom?', but also of *when* and *where* production would occur.

SOCIAL COMPLEXITY, DEMAND DIVERSITY AND EMPLOYMENT

Unemployment is highest in areas with simple infrastructures, a low degree of interdependence between types of activity, low participation rates in higher education, and limited demands for a range of goods and services. Employment rates are highest where there are complex infrastructures, a high degree of interdependence between activities, high participation in advanced education and a wide range of demand for diverse goods and services. The importance of social complexity and interdependence as employment generators can be illustrated first by comparing Melbourne's western suburbs (where unemployment levels are high) with the remainder of the city, and second by comparing Melbourne's west with the state of Tasmania.[28]

Melbourne has a population of 2·75 million, of whom 520 000 live on the basalt plain to the west, an area historically dominated by manufacturing industry, quarrying, meat processing, storage, dockyards, railway yards, airports and other service facilities. The industrial west is notably deficient in the 'people-intensive' activities which are central to post-industrial employment. It has only a small fraction of Melbourne's libraries, cinemas, restaurants, coffee shops, book, record, art or craft shops, government offices, tennis courts, swimming pools, sports fields, hospital beds or public transport. Few tertiary students, teachers, doctors, dentists, public servants, architects, musicians, artists, research workers, journalists or economists live in the region. Every profession in Melbourne is hierarchical, and the apex of each hierarchy is in the east.

The 'deprived west' is geographically part of metropolitan Melbourne, reads its newspapers, watches its television and listens to its radio stations – but psychologically the area is very distant from the city proper and the eastern suburbs. Access to Melbourne's facilities is strictly theoretical. There is a prevailing passivity in the west – it is *in* Melbourne, not *of* it – and its residents, despite the large population, generate little service employment in the metropolis. The social structure in the west is too simple, being based on the interaction of home → work → shopping → home. Some specific examples illustrate this cycle of non-access leading to non-participation. Eric Page's *Where to Eat in Melbourne 1979* (Pitman) reviews 700 restaurants in the metropolitan area: three of them are in the west. In June 1981, Melbourne had seventeen radio stations – none in the west (although 3WRB was expecting an FM licence in 1981). In the *Age Weekender* in a typical week (27 June–3 July 1980), of 646 metropolitan listings including films, dancing, jazz, folk music, art galleries, fairs and children's activities, only twenty-seven (4 per cent) were in the western suburbs. Of 550 courses offered in Melbourne by the Victorian Council of Adult Education, two were offered in the west. The east's residents are high cultural consumers, the west's are not: there is a psychological barrier in the west which inhibits participation in concerts, plays, film shows and library use.[29]

The importance of complex infrastructures as employment generators can also be demonstrated by comparing Melbourne's western suburbs with the less populous, but more diverse, society of Tasmania.

Tasmania is geographically isolated and dependent on its own social (but not financial) resources, with a population of only 413 000, 26 per cent fewer than Melbourne's west. Nevertheless, Tasmania has its own state government, a governor, a premier and a Cabinet of ten ministers, a parliament with fifty-four members, ten senators and five members of the House of Representatives, a Supreme Court, specifically Tasmanian professional career structures – teachers, doctors, architects, engineers, policemen, and so on – a substantial public service, trade unions, five television channels, fourteen radio stations, three daily and nine weekly newspapers, a symphony orchestra, a university, a college of advanced education (CAE), seventeen bookshops, museums, art galleries, two casinos, eighteen race courses, eleven airports, a large number of voluntary organizations, fourteen major hospitals, and over a hundred motels and restaurants. Compared to Tasmania's sixty-nine parliamentarians, Melbourne's west elects eighteen MPs (Commonwealth and state) and pro rata ought to be able to claim 1·3 senators as its own. It has barely 3 per cent of Melbourne's tertiary students (see Chapter 7 for further discussion of education participation rates).

If Melbourne's west had an autonomous regional government or became a state in its own right, confining the exercise of political and administrative power to local residents, this would immediately convert an impotent area into a strong political unit. A more complex society would emerge almost immediately, and generate much service employment. New sets of hierarchical power or career structures would be created, all with their apex within the west, and this would require an appropriate educational and social infrastructure. On the other hand, Tasmania would be an economic wasteland without sovereignty. It might be easier and quicker to proclaim the west's independence than to wait for the evolution of more rewarding and diverse work patterns through market forces. Service employment depends on an increasing complexity in social and cultural life, diversity of lifestyles, and a growth of transactions. Melbourne's west is crippled by social and cultural poverty more than by economic hardship. (Many of these observations are also true of regional and country areas in Victoria, although something has been done to provide cultural and tourist services.)

Sample comparisons of 100 adjacent shops chosen arbitrarily in the electorates of Lalor, Kooyong and Mallee indicate the divergent patterns of demand for goods and services, and suggest the range of potential employment in various areas.[30] (For profiles of these areas, see Chapter 3.)

	Lalor	*Kooyong*	*Mallee*
Food (basic)	23	5	18
Restaurants and/or gourmet take-away foods	3	6	4
Clothing and footwear	16	12	19
Furniture and home improvements	9	11	13
Real estate	8	5	–
Banks	6	1	11
Finance	4*	1	–
Gifts/souvenirs	4	1	1
Chemists	4	1	2
Health care (doctors/dentists, etc.)	1	9	4
Hairdressers/beauty care	2	11	7
Lawyers	3	1	1
Crafts/hobbies	1	10	2
Travel	1	2	1
Architects	–	2	–
Others	15	21	17
	100	100	100

* The heavy emphasis on real estate, banking and finance in St Albans reflects rapid increases in population in a relatively recently settled area with much new housing, whereas Hawthorn is an older suburb, with a high proportion of houses more than fifty years old, with long established families and a stable population.

The large number of basic food shops (e.g. grocers, butchers, bakers and greengrocers) in the Lalor and Mallee samples, and the small number in Kooyong, confirms the accuracy of Engel's Law: 'The proportion of income spent on food diminishes as income rises.'[31]

Increased complexity in transactions leads to a vast increase in the possible permutations and combinations of human activity, all of which are potential work generators (although the price will be higher inflation if the chain between production of goods and services is too long, as it tends to be in large cities). This leads to Jones' Fifth Law:

Rising levels of employment depend on increased demands for a diversity of services, many stimulated by education: simplicity of personal needs contributes to low levels of employment, and complexity to high levels. Over-specialization and economic dependence in particular regions on a single employment base (e.g. heavy industry or farming) inhibits the development of service activity.

The barriers to higher employment in working-class and rural areas are cultural, psychological and environmental rather than economic. Some regions with the highest productivity have the worst unemployment – it appears that high productivity and quality of life are inversely related. High growth rates, profitability, significant return on investment and reduction in waste are hallmarks of economic efficiency – but if they lead to high unemployment, demoralization and human wastage, then social losses may outweigh any economic gains. The elimination of telephone operators in isolated country regions was justified on economic grounds but represented a significant loss of social contact (and capacity to respond to emergencies). The 'efficiency' of highly specialized economies such as Manchester and Bradford have caused massive social disbenefits, because when labour input falls in established industries little new work evolves. By comparison, the less specialized, overlapping and apparently wasteful enterprises in Birmingham have been more socially efficient, with greater flexibility, new work creation and capacity to adapt.[32] Within metropolitan Melbourne, Lalor has greater economic efficiency than Kooyong, producing enormous 'value added' from its massive sophisticated industries – but its social efficiency is lower, and Kooyong has far more capacity to expand and diversify employment. Small businesses typically have higher unit costs than large ones: they are often less economically efficient, but more beneficial in social terms.

PRODUCTIVITY AND EMPLOYMENT

Productivity is a ratio:

$$\frac{\text{Output}}{\text{Input}}$$

Where tangible goods such as iron ore, coal, cars or shoes are produced 'output' is measured by their market value. The Production Index of Manufacturing Industries ranks Bolivia in first place for the world (1970:100; 1976:297), followed by Ecuador (1970:100; 1976:191), and then Brazil, Syria, Mongolia, Rumania, Malta, Turkey, Nicaragua and Senegal. On this basis Japan, the United States, the USSR, the UK, Switzerland, Sweden, Australia and other technologically advanced nations fall far behind the Dominican Republic, Iran, El Salvador and Ghana.[33] Bolivia comes first because tin has a high price and the miners' wages are very low. Ecuador and Senegal are high because they produce huge crops of bananas and peanuts and pay their workers starvation wages.

The annual growth rates for industry (AGRI) in the period 1970–76 were highest in Congo (22·6 per cent), South Yemen (17·7 per cent), South Korea (17·1 per cent) and Saudi Arabia (16·5 per cent).[34] Among the lowest-scoring nations in that period were Australia (1 per cent), the United States (0.9 per cent), United Kingdom (0.5 per cent) and West Germany (0.2 per cent). However, on the physical quality of life index (PQLI), which averages indices for life expectancy, infant mortality and literacy, of 151 nations, Bolivia ranked 92, Ecuador 71,

Brazil 70, Senegal 138, Congo 131, South Yemen 129, South Korea 55 and Saudi Arabia 118.[35] The highest-rating nations in the PQLI listing appeared to perform very poorly in AGRI:

	PQLI (151 countries)	*AGRI* (96 countries)
Sweden	1	78
Iceland	2	62
Netherlands	3	72
Norway	4	51
Denmark	5	85
Japan	6	55
Switzerland	7	n/a
Canada	8	56
France	9	73
United Kingdom	10	88
Australia	11	86

Does this mean that Australia should adopt Congo or Bolivia as an appropriate model for economic development? No, but it does suggest that current measurement of productivity is inappropriate, putting too much emphasis on the production of marketable commodities and ignoring quality-of-life factors. The Canadian GAMMA Report recommends the adoption of a QOL index in place of, or in addition to, GDP; the index to take account of thirty-seven factors including such areas as physical and psychological security, harmony with nature, self-actualization, and personal dignity. Unfortunately the report does not suggest how much weight should be given to each element, and does not attempt to rank nations.

The reason for the discrepancy is that, as discussed in Chapter 4, white-collar work performed outside the market economy is only measured by the cost of wages. Thus for teachers, output – the social and economic value of the work done – is priced at the value of input and its productivity in dollar terms does not rise or fall. The value of a miner or process worker, on the other hand, is calculated by the value of goods actually produced. Generally speaking, if seven workers out of ten in any given country spend their lives shovelling something valuable out of the ground, its labour productivity will be far higher than in a country where five workers out of ten are engaged in white-collar work outside the market economy (even if one of the ten operates a giant mechanized shovel). According to current techniques for measuring productivity, poets, philosophers, politicians, clergymen, social workers, public servants and academic economists have as market value only their salaries; while housewives have no economic significance at all. (Nor does the value of parks, beaches, clean air and clean water.)

In the 1970s and 1980s, employment growth in high-productivity, technologically based areas was low or negative in sophisticated economies, while employment in low-productivity areas which are not technologically based grew rapidly. The phenomenon of low productivity growth in some technologically advanced countries may seem surprising. It should not be so: the figures merely demonstrate that technologically advanced nations have heavy industrial bases with high literacy rates, which can lead to high employment growth in areas not producing tangible goods. Dr Peter Sheehan, of the Institute of Applied Economic and Social Research at Melbourne University, in an otherwise humane, sensible and lucid account of Australia's economic problems in the 1980s, wrote:

> The evidence is quite clear that technological change has not been a significant cause of the rise in *total* unemployment, and that it is unlikely to be so in the 1980s . . . [As] with real wages, the 'technology as a cause of unemployment' thesis implies an increase in labour productivity relative to what would otherwise have been the case . . . There has been no such exceptional rise in productivity, and if anything labour productivity has been lower than one might have expected over the past five years . . . In spite of all the talk about technology and its effects on jobs the dominant fact about [technologically advanced] countries has been the slowdown in productivity growth . . . [In] the United States, both employment and labour force participation have risen at rates unprecedented in the post-war period, and labour productivity per hour has increased by less than 1 per cent per annum.[36]

Dr Sheehan's analysis is based on several fallacies:

1 He aggregates employment figures, instead of disaggregating them so as to distinguish between employment loss possibly due to technological change and employment gains where technological change is irrelevant. For example, if twenty jobs are lost at a textile factory in Ballarat due to new technology, but the Ballarat College of Advanced Education creates departments of applied economics and Sanskrit and employs twenty more staff, then it can be argued that *total* unemployment in Ballarat is unchanged. Technically, this is true, but it is also meaningless and misleading because the people involved in the losses and gains are not exchangeable.

2 He disregards differing measures of productivity. If twenty retrenched Ballarat textile workers had earned a total of $200 000 per annum but produced between them textiles worth $600 000 – and if new technology enabled the factory's total previous output to be produced with a smaller labour input – then productivity would be significantly increased by the job reductions. If Ballarat's paid labour force were increased by twenty academics grossing $400 000 per

annum while the textile factory continued to produce the same value of goods as before, then the city's total productivity, as conventionally measured, would fall. If the textile factory closed down altogether, and more lecturers were employed than the total numbers of textile workers, there would be a growth in total employment *and* a fall in productivity in Ballarat. This is not a paradox, but an inevitable consequence of job displacement in high-volume productivity areas due to technological change and the steady growth of employment complementary to and not dependent on technology.

3 He fails to recognize that productivity is a ratio between outputs and inputs, and that it is increasingly difficult to have high employment, high participation rates, *and* high productivity. It is possible to have high productivity and low employment and participation, but we cannot divide a high number by another high number and obtain a high number as a result. This points up the inadequacy of current methods of economic measurement.

Another factor contributing to low industrial productivity growth per worker in rich countries is that the total value of output (e.g. in the motor industry) has to be divided by large numbers of people in the distribution chain, salesmen, advertising agents and other white-collar employees. If total employment in tertiary, quaternary and quinary sectors continues to grow, then the national productivity ratio will be maintained (but not increased) by the sheer volume of wage incomes which will continue to exceed manufacturing wage incomes and profits.

TWO MIRAGES: CUTS IN REAL WAGES AND THE RESOURCES BOOM

The Australian Treasury, and monetarist economists generally, have come to accept that Keynesian expansionist policies will not necessarily restore high levels of employment, and urge cuts in real wages. The British experience suggests this is not an answer. Real wages have fallen steadily throughout the 1970s (the resentment of many low-wage earners was reflected in the strong Tory vote in traditional working-class areas in the 1979 election), but employment has not recovered. It has been suggested, and confirmed by a computer, that a 3·21 per cent increase in all real consumption (i.e. private and government expenditure) plus a 6·15 per cent cut in real wages would lead to a 5 per cent increase in total employment.[37] This is a naive assumption, disregarding the actual tendency in industry over a long period to introduce low-cost technology rather than put on more employees. It also ignores complications related to overtime, and begs the question of how the wage cuts could be implemented.

Australia's resources boom is often quoted as the likeliest source of job creation, and the Minister for Employment and Youth Affairs (Ian Viner) has stated that each job in mining or smelting would have a multiplier effect of ten to one. The Department of Industry and Commerce has concluded that a multiplier of four is more likely. Aluminium smelting is expected to attract a total investment of $5600 million in the period 1981–85, leading to a creation of 6000 permanent jobs and another 24 000 indirectly.[38] Since the anticipated growth in people of labour-force age in the period 1981–85 is about 850 000, aluminium will contribute 0·7 per cent of the prospective jobs (at a capital cost of $933 000 per job), rising to 3·5 per cent if the multiplier of four is applied. It is hard to find evidence of comparable enthusiasm or publicity being generated over other resource-based industries, which will absorb a significant proportion of the additional 96·5 per cent offering for work. Australia is still groping for solutions.

7 Education and Employment

To expect a nation to be ignorant and free is to expect something which never has been and never can be.

Thomas Jefferson

Education, once a prerogative of the church and an instrument to help Bible-reading Christians ensure their salvation, became an attribute of citizenship during the secular changes of the Second Industrial Revolution. The granting of universal male suffrage in Britain, North America and much of Europe was accompanied in the period 1870–80 by universal, compulsory primary education under state or church auspices. The Australian colonies were among the first to adopt this.

Australia also shared in the educational revolution which followed World War II – universal secondary schooling and increased participation (largely by the middle class) in tertiary institutions. This trend, which continued in the United States, Canada, Japan, the USSR and much of Western Europe (although Britain was an exception) seemed to have run its course in Australia after 1975. At a time of increasing complexity and competitiveness in the world community, Australian education marked time – but as the advanced world was pushing ahead, this in fact meant a serious decline. In the 1980 OECD economic survey *Australia*, international comparisons were provided for many economic and social indicators. In the table of full-time school enrolments of young people aged 15–19 shown on the following page, Australia ranked fourteenth of twenty-three nations listed (1977).

The Australian figures conceal the truth: in working-class areas, the number of 15- to 19-year-olds in full-time education is about 15 per cent, in middle-class areas about 75 per cent. The figure of 45 per cent is indeed the average, but it distorts educational participation rates in specific areas. Internationally, Australia is ceasing to be intellectually competitive; internally, we are losing the essential preconditions for personal competence, social cohesion, employment prospects and the free flow of comprehensible information which makes democracy workable.

		Enrolment (%)
1	United States	73·7
2	Japan	70·9
3	Switzerland	70·1
4	Canada	64·9
5	Norway	63·6
6	Netherlands	62·7
7	Belgium	61·3*
8	Finland	60·8*
9	Denmark	57·4
10	Sweden	56·3†
11	France	54·6
12	Ireland	50·0†
13	Greece	45·4
14	Australia	45·0
15	New Zealand	44·8
16	United Kingdom	44·6†
17	Italy	43·9†
18	West Germany	41·5††
19	Spain	35·1†
20	Luxemburg	33.5*
21	Portugal	33·4†
22	Austria	32·0
23	Turkey	12·7*

* 1975 figures.

† 1976 figures.

†† East Germany's educational retention rates are not listed by OECD, but elsewhere it is ranked with the Netherlands. West Germany's low retention rate is misleading, reflecting the traditional dual system of vocational studies plus apprenticeship: more than 90 per cent of school leavers become apprentices and attend part-time vocational schools until the age of eighteen.

Education as a mirror of society: class distinction in Australian education

Australia has a split-level education system which perpetuates existing social, ethnic, class and regional divisions instead of eliminating them, which was the liberal hope for universal education. The first division produces the officers and NCOs of Australia's social army, and is:

1 predominantly urban middle-class;
2 achievement-orientated, with a high self-image;
3 aimed at producing qualifications for a career, presuming success, affluence, readily marketable skills, professional satisfaction, and personal autonomy in determining work patterns;
4 marked by a high participation rate in tertiary education.

The second division provides an educational minimum for the 'other ranks' and rejects from the system, and is:

1 predominantly working-class, ethnic, rural or regional;
2 failure-oriented, with a low self-image;[1]
3 aimed at producing competitors for a declining number of relatively semi-skilled or unskilled jobs, often poorly paid and with a low degree of work satisfaction and little personal autonomy;
4 marked by a low participation rate in tertiary education.

In *Poverty and Education in Australia*, Dr Ronald T. Fitzgerald concluded that:

> So long as access to careers is restricted to a minority of workers, the familiar stress on competition and academic success within the schooling system will combine to defeat all but very few children of low income families, irrespective of their intellectual ability. As a result, the growing gap between the haves and the have-nots in a so-called egalitarian society will continue to widen . . .
>
> Success in school and in the competition for rewarding careers is largely determined by such factors as social class, ethnic background and geographic location. The structural inequalities in our society are nowhere more evident than in our school systems. Far from being a way out for poor people, schools act as a sorting, streaming mechanism helping to maintain the existing distribution of status and power . . .
>
> People who are poor and disadvantaged are victims of a societal confidence trick. They have been encouraged to believe that a major goal of schooling is to increase equality while, in reality, schools reflect society's intention to maintain the present unequal distribution of status and power. Because the myth of equal opportunity has been so widely accepted by Australians, the nature of unequal outcomes has been largely ignored. Thus, failure to succeed in the competition is generally viewed as being the fault of the individual rather than as the inevitable result of the way our society is structured.[2]

Australia faces alarming increases in unemployment, and a widening social and economic division between the 'information-rich' and the 'information-poor'. It is essential to adopt a democratic, pluralist and egalitarian education system to give working-class, rural and migrant children the same range of educational options and quality of instructions – including access to tertiary education and the securing of professional qualifications – as are enjoyed at present by the urban middle and professional classes. More equitable access to education will not necessarily end the inequalities in society or lead to universal improvement in job status: it would deceive the poor to promise everyone better jobs as a result of more education. Many will achieve this, many will not – those who climb the socio-economic pyramid will continue to displace others and push them down. We must, however, assert that education is a good thing in itself – a human right and a consumer good – and that it is better to be educated and unemployed than uneducated and unemployed. In addition, education creates convivial-sector employment for teachers and reduces pressure on existing jobs by regarding students as 'workers'. Higher levels of education also create a wider range of needs to be met as a consequence of heightened perceptions, such as in travel, the arts, craftwork, books and hobbies – all of which are labour-intensive.

Australian participation rates in education are marked by enormous wastage among children of migrant families, or those living in the country, especially girls in the second division. The educational and occupational futures of most Australian children can still be predicted with a high degree of accuracy by asking only three questions: Where do you live? What school do you go to? What do your parents do?

In 1977, Australia's school population was divided into three groups:

	Pupils enrolled	*Enrolment (% of total)*
Government (state) schools	2 364 316	78·95
Catholic schools	502 044	16·76
Independent (private) schools	128 432	4·29
	2 994 792	100

The amount of wastage varies dramatically between the three school systems. The following are 1978 figures:[3]

	Government (%)	*Catholic* (%)	*Independent* (%)
Total school population (primary and secondary)	78·95	16·76	4·29
Retention rate past minimum school-leaving age	29·6	41·5	88·9
Wastage	70·4	58·5	11·1
School origins of first-year university students	58	22	20
School origins of first-year CAE students	65	22	11

Students from independent schools have disproportionately higher representation in the older universities (which have more prestige) in law, medicine and science, and in honours courses and post-graduate work generally, while students from government and Catholic schools (many of whom are intending to become teachers) are more likely to enrol for pass courses in arts and commerce. However, these figures understate the degree of bias *towards* the middle class and *against* the working class in securing entry to tertiary education. Government schools reflect the prevailing philosophy of their locations – and if it is taken for granted that a majority of pupils in a middle-class area will complete Year 12, enter a tertiary institution and work in a profession, then this will probably be a self-fulfilling prophecy. Similarly, in a working-class or rural area, if it is also taken for granted that most students will 'drop out' at the earliest opportunity and seek the declining number of available jobs, then this also will happen.

If 90 per cent of students in high school X (in a middle-class suburb) complete Year 12 and enter a tertiary institution, while only 10 per cent do so in high school Y (in a working-class suburb), obtaining a simple average from the figures would be seriously misleading about both schools. The conventional wisdom in Australia accepts the *embourgeoisment* of education almost without question. Middle-class politicians and educationists take for granted that it is futile to expect higher participation rates for the working class in education past Year 10 – and this view is widely accepted in the working class itself. The school system largely reflects middle-class values, and the teaching service has been an important means of providing professional training for young people from the working class. (Ross King of the Melbourne University Centre for Environmental Studies has calculated that there are 2500 teachers on the municipal roll for the middle-class city of Camberwell in Melbourne's eastern suburbs, and only 1600 in the entire western area of Melbourne, which has 19 per cent of the total population.) The implicit goal of the system is to enable students to complete Year 12 and enter a tertiary institution – and since this goal is not met by two-thirds of the pupils it is hardly surprising that formal education seems remote and irrelevant to so many. Students whose native language is Italian, Greek or Croatian find that their home culture is devalued – they must come to the school linguistically, the school does not come to them. Children whose parents have had little education, and who disliked it, have an inbuilt disadvantage. *Poverty and Education in Australia* describes the ways in which existing systems of education alienate poor people, and refers to surveys of early school leavers in the labour force:

> The tone of their descriptions of school was almost uniformly negative, sometimes hostile. They nearly all said they disliked school and saw it as a stage of life to be gone through and endured. They considered that schools operated against them by examining them constantly, judging them and failing them by criteria they found meaningless.[4]

Expectation of failure in education in particular areas leads to low levels of demand – and a self-fulfilling prophecy of educational failure leads to a continuing cycle of low educational participation.[5] The poor must have positive discrimination to help them break out of this cycle.

Education as a rationing device: 'credentialism'

Apart from the class-economic bias built into Australia's education system, there is a major paradox common to all systems in the Western world, unexamined and unresolved – that although most learning is achieved outside formal education, economic advancement and promotion are increasingly dependent on the possession of a 'credential'. We may think of economic and social activity as a game: once anyone could play, but as increasing numbers turn up to take part the organizers of the game demand to see a ticket.

In *De-Schooling Society* (1971) Ivan Illich made a radical critique of education. He argued that the great bulk of what we learn is picked up informally: that we assimilate far more between birth and the age of five than in any subsequent period at school (that the impact of television, for example, may provide in a few hours far more variety of sensory input than can be acquired in a year of formal tuition). We learn to speak informally and we do it fluently and unself-consciously: we learn to write formally, at school, and most do it awkwardly and reluctantly. Physical co-ordination, playing sport, swimming, driving, cooking, using tools, social relations (including sex, marriage and raising children), elementary reasoning, general patterns of thought and association – all of them central to our sense of identity and self-worth – are almost exclusively learnt outside the formal education system. However, the great bulk of what is thrust at pupils during their school years – grammar, formal logic, algebra, French, Australian history, music, art – creates a lifelong immunity. The learning curve, so dramatic in the first six years of life, cuts back in primary school, even further in secondary school, and by tertiary education it hardly deserves to be called a curve at all. However, as *Poverty and Education* found, 'education continues to provide the means of entry to the relatively few jobs that are prized by the community and rewarded with high status, high income and high job satisfaction' so that educational certificates are used as 'a rationing device'. The report endorses the words of Daniel Bell:

> Because the technocratic mode reduces social arrangements to the criterion of technological efficiency, it relies principally on credentials as a means of selecting individuals for places in the society. But credentials are mechanical at worst, or specific minimum achievement at best; they are an entry device into the system.[6]

In *The Credential Society* Randall Collins argued:

> The great majority of all jobs can be learned through practice by almost any literate person. The number of esoteric specialities 'requiring' unusually extensive training or skill is relatively small. Schools have relatively little effect

on learning, except as they . . . certify displays of middle class discipline. It has been by the use of educational credentials that the lucrative professions have closed their ranks and upgraded their salaries; and it has been in imitation of their methods that other occupations have 'professionalized'. We have elaborated a largely superfluous structure of more or less easy jobs, full of administrative makework and featherbedding, because technology allows it . . . In effect, leisure has been incorporated into the job itself.[7]

Professor Andrew Hacker commented:

Collins asks that we put aside our pieties about the complexity of modern knowledge. He is not arguing that we are equally talented, or that everyone can pursue every field. Rather his point is that most of the skills we use come to us on the job, and that what we learn in formal schooling has little effect on job performance. People who excel at what they do owe it to having temperaments suited to these tasks, aided by the trial and error which comes with everyday experience.[8]

We should recognize and pay for the concept of 'education for education's sake' rather than 'work for work's sake'. We are, after all, prepared to accept almost without question the costs of what Ralph Nader has called 'the involuntary sub-economy' – thousands of millions of dollars that consumers would not have paid if they knew or could control what they were getting – compulsive over-spending on private transport in the absence of viable public alternatives, and conspicuous consumption of clothes, gambling, and officially approved drugs of addiction for their own sake. We ought to be prepared to accept that the worth of a society can be measured not only by the consumption of goods and physical amenities, but also by its willingness to provide psychological amenities – knowledge, understanding, expanded consciousness, cultural responsiveness and increased creativity. We must recognize that psychological needs are at least as important as physical ones, and that education has always been (and will continue to be) our greatest industry in terms of the employment (meaning 'occupation', both paid and unpaid) of vast numbers of people.

Regional factors in educational participation

Education participation rates have an enormous cyclical influence in determining the type of community. Without a significant number of tertiary students, for example, there are unlikely to be many bookshops in an area. Without bookshops people are likely to devote free time to television, children are unlikely to be stimulated by books and may lose any desire to go on to higher education. They are less likely to participate in theatre or live music, and so on.

The electorates of Kooyong, Lalor and Mallee, already referred to

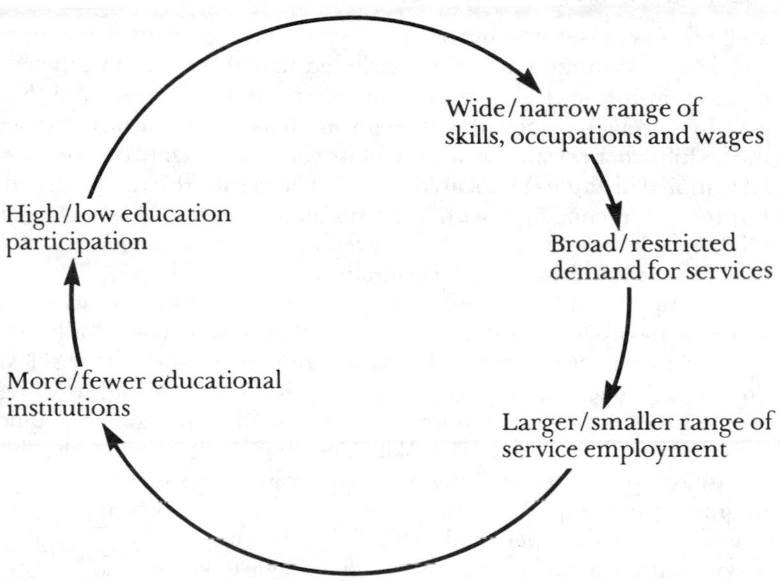

in Chapters 3 and 6, illustrate the diversity of access to education. The following figures have been calculated from the 1976 census.

	Kooyong	*Lalor*	*Mallee*
People with a degree or graduate diploma (%)	6·92	0·54	0·98
People studying at universities or CAEs (as % of those aged 17–24)	46·05	10·74	4·5

For every person with a degree or graduate diploma in Lalor there are two in Mallee and thirteen in Kooyong. The census figures do not distinguish between university and CAE students, between pass and honours students, or between full-time and part-time students: if full-time university students had been isolated, the ratio might have been more like one to ten. Lalor and Mallee have educational profiles like Mississippi, Kooyong like California. Lalor is part of an educationally disadvantaged area comprising 19 per cent of the metropolitan population but nearly a quarter of its children. Melbourne has three major universities (Melbourne, Monash, La Trobe), with a total enrolment of 38 562 students, and nineteen CAEs with 45 950 students.[9] (Deakin University at Geelong has 4400 students, some of whom come from

Melbourne's western suburbs. The report of the Victorian Fourth University Committee in 1971 concluded that there was insufficient demand in the west to support a university and recommended that priority be given to a fourth university in Melbourne's eastern region: this exhibited a certain circularity in its reasoning.) Only one tertiary educational institution is located in the western suburbs: the Footscray Institute of Technology, with 2580 students – only 3·05 per cent of Melbourne's enrolments. Mallee is even more disadvantaged because of its remoteness from tertiary institutions.

The crippling effects of educational and cultural poverty are both causes and consequences of unequal access to power. The middle and upper classes have access to information and power; the working class does not. While this disparity exists, it will be impossible to bring about a political redistribution of power and existing hegemonic structures must continue. The 'skill mix' shifts each year, to the disadvantage of the poor. Language ability was less relevant when migrants were employed as labourers or process workers in factories – but with higher personal skills being required in sophisticated service employment, a poor command of English will disqualify many migrants from work.

Percentage of 20–24 age group involved in tertiary education

	1960	*1965*	*1970*	*1973*	*1974*
Australia	13·13	16·04	16·62	20·13	22·17
Canada	16·04	26·35	34·59	35·08	34·69
France	7·45	14·21	16·09	—	18·04
Federal Republic of Germany	6·11	8·83	13·41	—	19·35
German Democratic Republic	16·27	19·87	32·81	—	24·50
Israel	10·19	20·04	20·03	23·32	23·54
Japan	9·45	12·90	17·02	—	22·14
Netherlands	13·20	16·75	19·51	22·25	23·49
Sweden	9·06	13·11	21·36	21·72	21·81
UK	8·50	11·95	14·37	16·22	—
US	32·07	40·18	49·43	51·53	53·61

Widening participation in further education

For Australia in the 1980s, to suggest that as many of our people are educable as are Canadians or Americans may seem a stunning proposal. It involves accepting the concept that working-class, rural or migrant children are educable, and revives the unfashionable doctrine of natural equality. In place of the hereditary principle in education, cultural consumption and sophisticated lifestyles (including choices of leisure activities), we should encourage people to develop self-confidence so that they can explore the universal cultural heritage for themselves – to find that education can be inclusive rather than exclusive; that *Hamlet*, *War and Peace* and the 'Jupiter' Symphony need not be the possession of a tiny elite; that is not compulsory to be passive members of a mass-exploitation market based on disposable pop culture. The table on p. 162, prepared by the Commonwealth Parliamentary Library, indicates the full- and part-time participation rates in tertiary education for selected countries. The US figure suggests that nearly all the population in the cited age group with an IQ above 100 (the mean) is involved in tertiary studies, either full- or part-time. In practice we could assume that many bright young people aged 20–24 would not be involved (e.g. women with children), and that there must be some with IQs in the 90s undergoing tertiary education. The Australian figures are misleadingly high because they include about 180 000 part-time students in Technical and Further Education (TAFE) undertaking vocational studies such as fitting and turning, food processing, hairdressing, dressmaking, sheet-metal working and similar courses which are excluded from the OECD classification of 'tertiary education'.[10] In the 15–19 age group, the estimated percentage distribution was as follows in August 1978:[11]

At school	38.8
Employed	43·8
Unemployed	9·8
Tertiary education	7·6

In the overlapping 17–22 age group, the percentage figures for participation in tertiary education in 1977 were:[12]

University	
Full-time	5·4
Part-time	0·7
TOTAL	6·2
CAEs	
Full-time	4·5
Part-time	1·0
TOTAL	5·5
TAFE	
Full-time	1·8
TOTAL	1·8

In terms of full-time enrolments in tertiary institutions in the 17–22 age group, some comparative percentages are Australia, 8·1 per cent

(1976); Canada, 13·1 per cent (1970); Japan, 16·8 per cent (1976); France, 9·8 per cent (1976); United States, 23·4 per cent (1976).[13] Examination of the comparative extent of educational absorption in tertiary education in various countries suggests that young people, whether 15–24 or 17–22 years, may be arbitrarily classified into three groups:

Group 1. Young people of superior ability, or those of above-average ability who receive the advantage of superior teaching and family encouragement, who wish to undertake tertiary education, actually do so, and generally succeed. In most, but not all, cases they enter professional employment. Most come from upper- and middle-class backgrounds.

Group 2. Young people with the capacity to undertake tertiary education and to succeed, but who for a variety of reasons have come to accept self-limitation and prefer to seek early entry into employment, early marriage and a home and family rather than postponing those goals to compete for a degree and then seek work in an uncertain future. Most drop out between Years 10 and 12, others complete Year 12 and then end their studies. Many in this group, especially those from working-class backgrounds, may have been socially conditioned by the family, the school and the environment to accept what is, in fact, an unrealistically low level of aspiration. This group may also include people of superior ability whose capacity has never been properly assessed.

Group 3. Young people who have no intention of going on to tertiary education and would be unlikely to succeed. These generally drop out of secondary school at the minimum leaving age (or even before if exemptions are granted). Group 3 may include some young people of superior ability who are grossly disadvantaged by school. This group is predominant in the working class, in rural areas and amongst migrant families.

In comparing Australian participation rates with other countries, we may conservatively estimate the size of the three groups as follows:

Group 1	20%
Group 2	40%
Group 3	40%

In a normal distribution curve, 20 per cent of the community would have an IQ above 113·6, 60 per cent between 86·4 and 113·6, and 20 per cent below. The following factors should be taken into account:

1 High drop-out rate (60 per cent) between Years 10 and 12.
2 Differing retention rates between government, Catholic and independent schools.

3 Variation in retention rates at Year 12 between states (30·9% in NSW, 23·1% in Victoria).
4 Declining proportion of students proceeding from Year 12 to higher education (54·5% in 1974, 48·7% in 1977).
5 Extent of regional (i.e. urban versus rural) variation in educational participation.

A 1973 study of Australian Jewish youth, by R. Taft, indicated that 60 per cent expected to undergo tertiary education and that 40 per cent expected to gain a degree. In the ACT, 69 per cent of government-school students stay on to Year 12, compared to a national average of 35 per cent. Of all migrant groups, German children have an atypically high participation rate in tertiary education.

It is striking that in the 15–19 age group, comprising 1 248 900 young people, the unemployed are more numerous (9·8 per cent) than those engaged in tertiary education (7·6 per cent), and that the numbers of school leavers going on to higher education is declining. In August 1978, 17·5 per cent of the 15–19 age group *in the labour force* (i.e. those not at school or in tertiary education) were unemployed, compared to a 4·8 per cent rate for those aged twenty and over. In October 1976, the Fraser government appointed a committee of inquiry into education and training, under the chairmanship of Professor Sir Bruce Williams, to review possible social and economic developments to the year 2000 as they related to education. The Williams Report confirmed that 'while 60 percent of our 15–19 age group are in the labour force, comparable figures are 24 per cent in Japan and 28 per cent in the United States of America'.[14] These figures should help dispel two prevailing myths – first, that Australia has an atypically low number of young people in work. The second myth is that Australia has too many young people in tertiary education: in fact, the number is unusually low relative to nations of comparable economic development. Australia's future economic and social growth will depend on increasing rates of educational participation at the higher levels.

With an estimated 1 475 000 people in the 17–22 age group, Australia has 11·7 per cent in full-time education (172 600). If we had the full-time participation rates of the United States, Japan, or Canada, these figures would rise dramatically. In order to overcome the twin problems of high youth unemployment and low participation rates in tertiary education, it is essential to increase the numbers of people in Group 2 in universities and CAEs. In the United States, Group 1 and Group 2 compete with each other in tertiary institutions, leaving many comparatively routine jobs available for people in Group 3. In Australia, Group 1 proceeds through tertiary education unchallenged by Group 2. Because the people in Group 2 are apprehensive about their economic future, they leave school in Year 11 or Year 12 and compete with Group 3 for the available jobs: they get them, leaving Group 3 to face chronic unemployment.

The quality of Australian education

In the 1960s, there was a strong public conviction that Australian education was not good enough, that new schools had to be built and the status and salaries of teachers raised. During the Whitlam Labor government, Commonwealth expenditure on education rose from $1774 million in 1972–73 to $4082 million in 1976–77, an increase of 230 per cent. After the Whitlam government fell, there was a political reaction against continued massive increases in educational expenditure – a sense that education had had its turn, and that more money had to be spent on roads and hospitals instead.[15] This view is mistaken.

The contracting birth rate since 1950 – falling from 23·4 live births per 1000 mean population each year to 16·1 in 1977 – means that the number of people aged nineteen or less in 1978 was a smaller proportion of the population than at any time since 1954 (35·2 per cent). (The figure peaked at 38·4 per cent in 1966, and is estimated to fall to 30·6 per cent by the year 2000.) If current absorption rates are maintained, this will mean fewer teachers and fewer pupils. The middle class benefited, at least in part, from an educational revolution in the period 1950–75, but it ended without having touched the working class, migrants and Aborigines. A further revolution in rising levels of expectation is needed now. To decide otherwise will be to perpetuate a criminal waste of human capacity, to deprive the poor of a major capital asset – access to education – and sentence them to permanent inferiority. This is a time to push ahead in education, not to pull back. The 1970s represented a turning-point in the history of education – for the first time in two hundred years of educational statistics, participation rates began to fall, not just in Australia but in much of the technologically advanced world.

Relieving youth unemployment is the greatest single social priority facing Australia. As outlined in the last section, youth unemployment is compounded by three factors:

1 falling educational participation rates at higher levels;
2 the 'displacement' effect, where those already stigmatized by failure at school lose out in competition for routine jobs against people who have the capacity to undertake tertiary studies (but do not);

3 very low working-class participation in higher education.

The Williams Report was presented to Parliament in March 1979, debated briefly and then adjourned. (The 31st Parliament ended without any further examination of it.) The report is fundamentally disappointing (although far superior to the Myers Report, its exact contemporary): it takes an ultra-cautious view of the impact of technological change, and assumes that the next twenty years will be a mirror image of the previous twenty. The only values it espouses are economic and materialist – it makes no attempt to advance an educational philosophy which, rightly or wrongly, could have been a catalyst for debate: instead it raises a number of additional questions which it hopes will be examined by the community at large. If a highly expert group after twenty-eight months was unable to come up with specific policies, it is hard to see how a dispersed and non-expert community could do it. The report provides valuable state and systemic statistics, but there is no class or regional analysis of educational participation at all, apart from a casual reference to the Tasmanian Educational Next Decade (TEND) Committee Report. An October 1975 survey of technical college entrants in Tasmania showed that one-half had weaknesses in the area of literacy and numeracy, while a New South Wales Education Department inquiry found that six per cent of 15-year-olds were so deficient in their capacity to calculate that they would be required to lead lives that would enable them to avoid calculations that the normal adult could expect to encounter in the course of everyday living. A further 25 per cent had 'demonstrated mastery of less than the full range of basic, elementary calculating skills'. The NSW Inquiry also found that 17 per cent of 15-year-olds were not able to read or could only read at a 'level below satisfactory comprehension of simple passages'. A further 32 per cent were reading at levels of comprehension below the 'competence required for a fully literate adult life'.[16] The Williams Report endorses the 1977 finding of the Australian Council for Educational Research that about 25 per cent of school leavers lack sufficient competence in reading, writing and numeration.[17] However, it agrees that schools are not responsible for prevailing levels of youth unemployment and should not be made scapegoats (as is fashionable with many employers and conservative politicians). This is correct: the school system, the Williams Report, and Parliament are no more mediocre and self-satisfied than the nation itself. No single element should be blamed.

The Myers Report expresses alarm at 'the small proportion of students who take technologically oriented subjects in the last years of secondary education' and urges more mathematics training.[18] This reflects the often-expressed concern in the business world that too many students are undertaking 'soft' subjects such as literature, philosophy and political science rather than 'hard' ones such as mathematics, physics and chemistry. This reflects the naive view that exact sciences are rigorous and studies based on value systems are not. No doubt the humanities are taught in an insufficiently rigorous way in Australia and teacher training has much to answer for, but Australia's future will depend as much on its politicians, writers, artists and humanities teachers as on its engineers and chemists. Our primary emphasis in education ought still to be on the general rather than the specific and vocational. It is no coincidence that Sweden and Switzerland, the most scientifically innovative of all nations per capita, had the highest literacy rates in Europe two hundred years ago. Their excellence springs from rigorous levels of *general* education and culture: if their educational goals had been set only to meet technological objectives, they would have failed. (The significant role played by Scots in British economic history was based on Scotland's high level of literacy.)

As Australia moves into an information-based post-service society, the greatest hope – perhaps the only hope – for a democratic and egalitarian community will be to affect a further revolution in education. The gap between middle-class and working-class expectations and performance in education, if left to continue, will perpetuate two societies within the nation. Australia is essentially a 'deference' society in which a large proportion of the poor and the badly educated have such a low self-image that they readily accept their incapacity for higher education or skilled work.

For most people school is a tedious once-in-a-lifetime experience to be endured before going to work, and success or failure there largely determines what kind of job (if any) is secured and how much is earned. Education aimed at tertiary professional qualifications will seem utterly remote to a significant number of young people and their families. A person born, for example, in 1965 may well live to the year 2050: it is pointless to provide him with a basic education which is designed for the economic world of 1980, when that world will change out of recognition by the end of the decade. As Marshall McLuhan says in *The Medium is the Massage*, 'Today's television child is bewildered when he enters the nineteenth-century environment that characterises the educational establishment.'

The school is an excellent sorting machine; it classifies people and sets them apart. If a person attends a poor school and lives in a poverty-stricken social environment, the school system will do relatively little to overcome those disadvantages – it may in fact extend them through life. Cultural poverty seems to be inherited, and is extremely difficult to shake off. It is an irony that as we approach the goal of a satisfactory class-pupil ratio in conventionally equipped classrooms – an end that has been pursued by all parties for some time – this may recede, turn into a mirage and prove irrelevant for the educational needs of a post-service society.

Formal education is a form of social Darwinism in which a minority flourish and the majority drop out. We have readily accepted the elitist concept that only about 10 per cent of young people are capable of benefiting from tertiary education and that only 30 per cent should complete secondary school. The poor are the white Negroes of society, but if they *were* black their degree of disadvantage would be recognized as a national problem and massive action taken to overcome it.

Education for outer and inner life

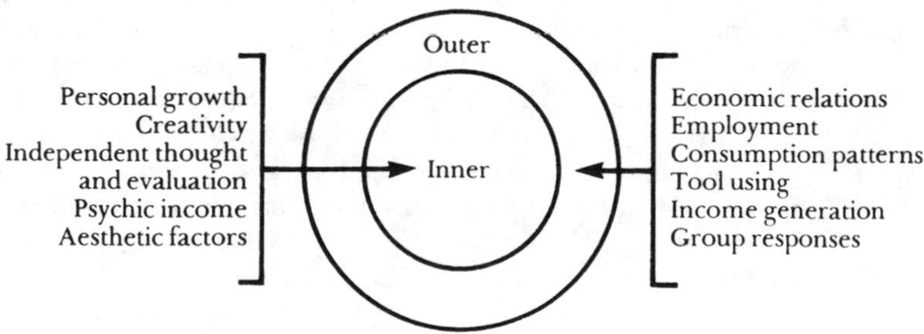

Middle-class – first division – education puts a heavy emphasis on personal development, and working-class – second division – on qualifications for income earning. Emphasis on personal development in middle-class education in teaching languages, the arts and literature does not inhibit employment prospects or the capacity to generate income: quite the reverse.

While consumption is the standard measurement of outer life on which we put so much emphasis, creativity or self-expression remains the key to inner life. The increasing tendency towards division of labour and fragmentation of knowledge forces many people to seek fulfilment outside work – by active participation in sport, gardening, hobbies or the arts – or to suppress self-expression through increased dependence on alcohol, drugs, spectator sport and television. As James Coleman argued, life used to be experience-rich and stimulus-poor; now it is experience-poor and stimulus-rich. 'Culture' is often perceived as an optional extra, reserved for the highly educated: the icing on the cake. Four thousand years of human history before the Industrial Revolution demonstrate how creativity and work were integrated and that the urge to make appears to be universal. Some nations have had long traditions of intense personal participation in specific cultural forms, e.g. pottery, calligraphy and poetry in Japan, music in Hungary, and soccer in Latin America. In addition, creativity has often been associated with a happy and vigorous old age.[19]

The 1970s were marked by increased activity in outdoor recreations: playing conventional sports and games rather than watching them, especially walking, climbing, birdwatching, camping, orienteering, canoeing, sailing, swimming – all of them active, adventurous, self-chosen and mostly environmentally responsible activities. But the degree of activity was closely related to class and educational background – boating and birdwatching were not major occupations of the poorly educated, urban working class, who were victims of cultural poverty in this area as well. In working-class education, highly specific training (e.g. in typing and shorthand, dressmaking, cooking, wood- and metal-working) does not necessarily guarantee employment, income or personal development: again, the reverse is true. The idea that a young man's prospects could be ensured by putting a spanner in his hand, pushing him out the door and watching him go on to fame and fortune is now anachronistic, dating from the time when people were described, by their function, as 'hands'.

The extent of unrecognized and untapped talent in the working class remains the great unknown in Australian education.[20] It is disguised to some extent by IQ testing, which appears to be an objective endorsement of what is instinctively believed by many teachers, educational administrators and politicians – i.e. that the working class is largely ineducable at higher levels (and by inference, that the middle class are intellectually superior). IQ testing certainly identifies people who scored well or badly in IQ tests, but it is a heroic assumption to assert that IQ scores and personal capacity are synonymous. The testing procedures themselves – and the testers – have a significant middle-class bias. In Britain and the United States middle-class children tend to score 10-15 points higher, across the whole range, than their working-class contemporaries. This is not to deny that there is a range of intelligence in every class and community, but a dull middle-class child may well outscore an alert working-class or non-English-speaking child. Failure at school is taken as an official certification of incompetence.

The Williams Report made it clear that there is a yawning gap between those who receive tertiary training and graduate to professional employment, and those who leave school with minimal and unmarketable skills appropriate for the contracting number of routine and repetitive jobs. The question is: How can working-class people be given access to education which maximizes their skills and leads to satisfactory personal and vocation outcomes?

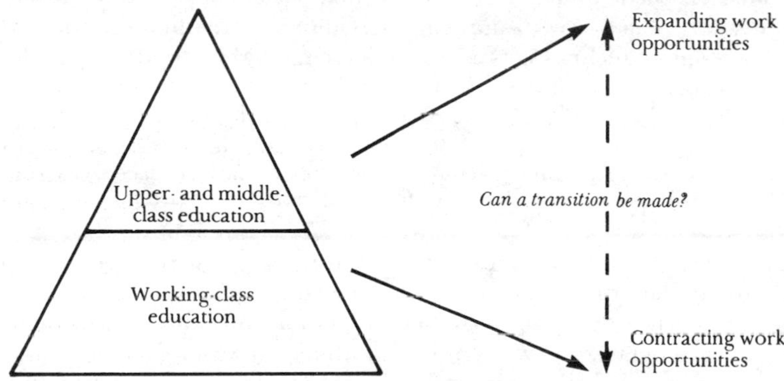

The main relationship the working class has with tertiary education is financial. Taxes from working people are used to subsidize university education for the children of the affluent, while working-class children are effectively excluded. (Abolition of tertiary fees by the Whitlam government has provided major benefits for mature-age students: it has not attracted more school leavers.) There is no such thing as 'too much education', although the affluent often claim that the children of the poor have been 'over-educated' (they rarely say it of their own). Many conservatives reject and deride the notion that the poor should be encouraged to pursue higher education, but there is also a 'radical chic' position which argues, perversely, that tertiary education is a bourgeois trap designed to destroy working-class solidarity, and to discredit proletarian culture and value systems: 'It is condescending to try changing them: leave them alone.' (Presumably access to superior dentistry, better nutrition, longer life expectancy and a wider range of interests and activities would have similar demoralizing effects.)

Learning is a process of growth, self-actualization and self-recognition – a means of pursuing the abundant life, assisting people to understand the world around them and the world within, to enlarge their personal ranges of choice, experiencing the satisfaction of creation and/or understanding. We should promote the concept of recurrent education as the essential right of every person as a means of self-development.[21] This is particularly serious in an age of technological revolution. If a lifetime's professionalism can be acquired, replicated and disseminated in a few seconds by computerized technology, then what value can be put on human experience in the workplace? In a period of increasing emphasis on high technology, it is essential that our education promotes humane and pluralist values, and strengthens individuals *vis-à-vis* their social environment. In the prevailing philosophy of machine worship, it is assumed that people must adapt to technology rather than technology adapting to people. As R. P. Blackmur wrote in 1954:

> The crisis of our culture rises from the false belief that our society requires only enough mind to create and tend the machines together with enough of the new illiteracy for other machines – those of our mass media – to exploit. This is perhaps the form of society most expensive and wasteful in human talent mankind has yet thrown off.[22]

The strength of the book-based culture was its relationship to personal autonomy, making it possible to absorb material at one's own rate – to stop, to go back, to ponder, note, brood and reconsider. In an era of mass, one-way communication, where language is persistently devalued, this view is made to seem increasingly anachronistic.

In conclusion, we need to avoid two polar extremes in personal or social development – the world of privatized experience, and the technologically determined society. The first is essentially fragmented, eclectic, self-absorbed and self-referential: the significance of the political process is trivialized and replaced by the 'therapeutic outlook', incorporating drug dependence, 'dropping out' and 'turning off'. In the second case, objectives are externally set: individuals are required to fit in to social goals, the prevailing technology, and economic norms. Cultural needs are met by coloured TV, newspapers and spectator sport – i.e. mass-produced commodities. There is an almost total emphasis on outer life and consumption levels, and rejection of the inner life. It is necessary to square the circle – to attempt some compromise between the inner and outer life – to develop individual personality in a cultural context and then to operate within a broader social, political and economic framework.[23]

8 The Information Explosion and Its Threats

A popular Government, without popular information, or the means of acquiring it, is but a prologue to a farce or a tragedy, or perhaps both. Knowledge will forever govern ignorance and a people who mean to be their own governors must arm themselves with the power knowledge gives.

James Madison

The fragmentation of knowledge

Australia is an information society in which more people are employed in collecting, storing, retrieving, amending, and disseminating data than are producing food, fibres and minerals, and manufacturing products. In the United States, almost half of the paid labour force is engaged in the information sector of the economy. The information society is marked by a shift away from employment in producing goods and services towards that of services and information, and an unprecedentedly rapid increase in the volume of readily accessible knowledge, often called 'the information explosion', a phenomenon which poses social and political problems. Access to knowledge, capital or wealth is roughly equivalent and there is a widening gap between the information-rich and the information-poor whereby the unskilled become an intellectual proletariat. The problem of control in an information society is largely unrecognized and undiscussed, and yet it raises the 'Who/Whom?' question which Lenin described as the basis of all political debate: 'Who does what to whom?' Is access to information to be centralized and subject to monopolist or oligopolist control, or is it to be dispersed, decentralized and widely available?

The sheer volume of available information and increasing emphasis on 'reductionism' – the concept that complex ideas or systems can be completely understood by analyzing their components[1] – has led to the fragmentation of knowledge. No single person can be expected to understand the technical significance of every major development in science and government – the time for adequate study is not available. The democratic system may become increasingly irrelevant as a means of determining and implementing social goals, or allocating funds on a basis of community needs, if elected persons do not understand how to evaluate and relate the segments of information in which each expert works. The expert relies on detailed but partial knowledge. The generalist relies on what he is told. Who links the experts?

The fragmentation of knowledge may lead to an incapacity or unwillingness to examine technical questions in a wide social context, with a failure to connect and, overall, an inability to comprehend what is going on. This is likely to lead to a loss of power by democratic institutions, and to increase the power of strategically placed minority groups occupying the commanding heights in particular areas of society – technocrats, public servants, corporations, unions.

The Indian fable of the four blind men and the elephant illustrates the problem of reductionism and the fragmentation of knowledge. The first man hugs one of its legs and says 'An elephant is like a tree'; another grasps the tail and says 'No, it is like a rope'; a third holds an ear and says 'It is like a sail'; and a fourth grips the trunk and says 'It is like a hosepipe'. All have expert knowledge, none will defer to the other, and yet the whole is somewhat more than the sum of the parts.

We are all the victims of the fragmentation of knowledge in which 'specialists of various stripes are left to trade on each other's ignorance . . . In the complexity of this world, people are confronted with extraordinary events and functions that are literally unintelligible to them.'[2] Effective decisions on technical matters – even where policy becomes involved – are made at comparatively junior levels in bureaucratic hierarchies. The initiating officer, who is expert in his own area but may be naive and inexpert outside it, prepares a memorandum; his permanent head, who does not understand it but feels that the recommendation must have been adequately monitored and checked out in the department, passes it on to the minister; the minister does not understand it either, but feels that is must be correct if the department endorses it and passes it on to the Cabinet; it is then endorsed by Cabinet, none of whom understands it, and becomes operational, either through ministerial fiat or by legislation: in the last case it is endorsed by the peoples' representatives in parliament, none of whom understands it either. But if challenged, the junior officer would say: 'It wasn't my decision. I was asked for a technical recommendation and I made it – my responsibility ended there.' There is a long series of people with pointing fingers, all saying 'It was his decision, really . . .'

'DISJOINTED INCREMENTALISM'

As Charles Lindblom has noted, governments operate on a philosophy of 'disjointed incrementalism' where societies become too large and complex for decision-makers to have a synoptic vision of all factors relevant for policy-making, and that looking for an all-encompassing overview 'assumes intellectual capacities and sources of information that men simply do not possess'.[3] This raises political and philosophical problems which are rarely mentioned in parliament and never discussed at elections. Specialists (such as public servants who are not directly accountable to the community) take power from generalists (such as ministers or MPs, who *are* accountable) because the specialists understand their own particular area of expertise – but not the whole – while the political generalists do *not* understand the parts and are gradually losing their grip on the whole. The Westminister concept of ministerial responsibility evolved when government operations were comparatively simple and a diligent minister could comprehend the full range of activities in his department. In 1859, when Lord Palmerston was re-elected as Prime Minister for the last time, he told his colleagues that the range of subjects which could be legislated on would be exhausted within the life of that parliament (seven years). Palmerston's judgment has not been vindicated.

The Merovingian kings in eighth-century France have had a very bad historical press because they lost their power to lesser officials called the 'mayors of the palace'. The sovereigns retained their positions of honour, but were called *les rois fainéants* ('the do-nothing kings'). This is an exact parallel to the position that parliamentarians find themselves in *vis-à-vis* the public service. In specific area, public servants are information-rich and their ministers are information-poor. Compared to voters, MPs are information-rich: compared to the bureaucracy they are information-poor. The increasing diversity and ubiquity of government operations has led to an inexorable growth in the public service, and decline in ministerial accountability. This can be illustrated by comparing operations in the Australian House of Representatives for 1901 and 1980.

In 1901 – the year the Commonwealth of Australia was established and the first elections were held – the House of Representatives had seventy-five members, there was a population of 3 774 000, a public service numbering 11 191, and there were total outlays of £4 024 000 for the financial year 1901–02. (Allowing for a deflator factor of eighteen and converting to dollars, this was the equivalent of $144 864 000 in 1980 currency – or $38 per head of the 1901 population). In 1901, the House of Representatives sat for 113 days, a total of 866 hours (equal to just over eleven and a half hours per member).

In 1980, also an election year, the House of Representatives had 124 members, there was a population of 14·4 million, public servants, employees of statutory corporations (quangos) and members of the armed forces numbered 469 100, and there were total outlays of $39 000 million for the financial year 1980–81 (or $2708 per head).[4] In 1980, the House of Representatives sat for fifty-one days, a total of 496 hours (equal to four hours per member).

Between 1901 and 1980, per capita public expenditure in Australia increased by a factor of seventy-one, after adjusting for constant money values, but the total time for scrutiny and debate in parliament has fallen to little more than half. A decreasing proportion of outlays by the Australian government was authorized by the annual budget and brought specifically to the Parliament for review. In the financial year 1979–80, 67·8 per cent of outlays were provided for by 'permanent and special appropriations' made under 148 different acts (which are not reviewed annually) and only 31·2 per cent in the appropriation bills (which are).[5]

Like nature, administration abhors a vacuum, and the decline of parliament as an effective instrument has been accompanied by a rise in the power of corporate elements – public service, mass media, trade unions, and industry generally. The public service has continual access to EDP facilities – members of parliament do not, except as provided 'by informal arrangements . . . on a courtesy basis'.[6] A *National Times* profile of John Stone, the Secretary of the Treasury, concluded with these words: 'The Treasury runs a relatively sophisticated model of the economy that contains nearly 150 mathematical equations [that] no politician could hope to understand.[7] That is probably true – but if we act on the assumption that no politician is equipped to propose an alternative view to Treasury, what future is there for the democratic political process?

The community is the collective victim of profoundly unequal access to information. MPs are like members of a football team which never plays at home – the public servants have collectively about 85 per cent of the information and MPs have about 15 per cent, mostly from leaks or newspaper reports. Even politicians are limited to the 24-hour day (sometimes less), and by their individual capacity to absorb material. Generalists are elected by other generalists to represent them, but are hopelessly outnumbered and outmanoeuvred by the specialists who are experts in their own tiny area but who often do not have (or claim to have) any broad grasp of society's problems.
I propose Jones' Sixth Law:

The amount of time spent by generalists in making technically based decisions is in inverse proportion to the complexity of the subject matter.

In an increasingly complex and sophisticated technical age, in which more and more voters understand less and less about what is going on around them, the role of democratic electoral processes may become irrelevant. A parliament composed largely of generalists may find itself in the hands of experts or technical specialists who will invite it to rubber-stamp what they are doing. In time they will not even do this: post-industrial society in Australia may lead to rule by technocrats – a situation described in the United States as 'techno-urban fascism' or 'friendly fascism'.[8] This must be resisted at all costs. Control of information resources may lead to excessive and disproportionate exercise of power by the strong over the weak. Individual freedom and individual efficiency will largely be determined by each person's ability to secure the right information at the right time.

The saga of Melbourne's West Gate Bridge is one example of major decisions, committing scores of millions of dollars, being made not only without the knowledge of the government or parliament but without understanding by sections of the supervising authority itself. During construction of this bridge across the Lower Yarra in October 1970, a span collapsed and thirty-five men were killed. The original Freeman Fox design was criticized by a Royal Commission and abandoned, and a modified plan devised by Dr W. A. Fairhurst. In 1972, the Fairhurst plan was scrapped by the West Gate Bridge Authority and replaced by a radical new design by Karlheinz Roik and Hans Wolfram which added $80 million and five years to cost and time estimates. The Victorian government was informed neither of the basic design change nor Fairhurst's dismissal, and the responsible minister (R. J. Hamer) defended the Authority's new design on the basis that Fairhurst had devised it. (In fact he bitterly opposed it.) Later ministers justified the design on the grounds that a four-man committee of engineering professors had endorsed it. The committee never met: each member was shown a segment of the bridge documentation falling within his expertise, and asked to give an endorsement.

A second illustration of the point is the Melbourne Underground Rail Loop Authority, which concentrated on the technical problems of tunneling through impervious rock in the city's centre without considering whether there was any provision for an increased catchment area for rail passengers. Two further examples in my experience include Mandata, and 'notional overpayments' to chemists under the Pharmaceutical Benefits Scheme.[9]

KNOWLEDGE AS A FACTOR OF PRODUCTION

In 'The Poverty of Philosophy' (1846), Marx wrote:

> Social relations are closely bound up with productive forces. In acquiring new

> productive forces men change their mode of production; and in changing their mode of production, in changing the way of earning their living, they change all their social relations. The hand-mill gives you society with the feudal lord; the steam-mill society with the industrial capitalist.

We now should add: 'The computer gives you society with a range of options – the concentration of knowledge power in a few hands or its dispersion throughout the community.' Information is a vital national, community and personal resource – a form of capital. 'Knowledge is power', and access to information is analogous to access to power.

In 1962, Fritz Machlup argued that there were essentially five types of knowledge:

1 Practical.
2 Intellectual.
3 Small-talk and pastime.
4 Spiritual.
5 Unwanted ('outside his interests, usually accidentally acquired, aimlessly retained').[10]

He also isolated those economic activities which primarily involved the manipulation of information – education, research and development, printing and publishing, photography and sound recording, stage and screen, radio and television, advertising, telecommunications, conventions, information machines (computers, printing devices, office equipment), professional, financial, business and government services – and estimated that in 1958 'knowledge industries' accounted for between 23 and 29 per cent of US gross national product. He thought they would grow at a rate of about 10 per cent annually: this has been true of their turnover, but as employers they levelled off in the 1970s.

In *The Age of Discontinuity* (1969), Peter Drucker adopted Machlup's analysis and referred to the 'knowledge economy':

> From an economy of goods, which America was as recently as World War II, we have changed into a knowledge economy . . . The statistics, impressive though they are, do not reveal the important thing. What matters is that knowledge has become the central 'factor of production' in an advanced, developed economy. . . . Knowledge has actually become the 'primary' industry, the industry that supplies to the economy the essential and central resources of production. The economic history of the last hundred years in the advanced and developed countries could be called 'from agriculture to knowledge'.

Drucker noted that the development of computers would intensify the growth of the knowledge economy. 'The computer is to the information industry roughly what the central power station is to the electrical industry . . . Information, like electricity, is a form of energy.' In 'The

Economics of Knowledge and the Knowledge of Economics', Professor Kenneth Boulding wrote:

The second law of thermodynamics informs us there is constant degradation and decay . . . and all processes in time are seen merely as the exhaustion of pre-existing potential. It is only information and knowledge processes which in any sense get out from under the iron laws of conservation and decay . . . We then see any developmental process as a combination of rote knowledge . . . and new knowledge . . . The recognition that development . . . is essentially a knowledge process has been slowly penetrating the minds of economists.[11]

Daniel Bell's concept of 'post-industrialism' was based on what he called the 'axial principle' of the 'centrality and codification of theoretical knowledge', in which the primary resource was human capital and the 'economic ground' was science-based industries. 'What counts is not raw muscle power, or energy, but information.'

The information explosion

Daniel Bell defined 'information explosion' in the following way:

A reciprocal relation between the expansion of science, the hitching of that science to a new technology, the growing demand for news, entertainment and instrumental knowledge, all in a context of a rapidly increasing population (of greater literacy and more schooling), of a vastly enlarged world that is now tied together, almost in 'real time' (i.e. simultaneously), by cable, telephone and international satellite, that is made aware of each other by vivid pictorial imagery of television and which has increasingly available, on national and international bases, large data banks of computerised information.[12]

The term, although widely used, is a misnomer: an explosion is a sudden violent event which quickly subsides, but the dramatic expansion of available knowledge seems likely to continue indefinitely. This is suggested in *The Computerized Society* (Penguin, 1973), where James Martin and Adrian Norman argue that

the sum total of human knowledge changed very slowly prior to the relatively recent beginnings of scientific thought. It has been estimated that by 1800 it was doubling every 50 years; by 1950, doubling every 10 years; and that presently it is doubling every five years. One wonders how long this acceleration can continue. Computer technology may make a frighteningly high rate of increase possible for centuries.

In 1973 Professor Georges Anderla concluded, after examining 'Delphi' surveys (an evaluation technique developed by the Rand Corporation) of several hundred scientists:

Before the end of the decade 1970–1980 all the essential conditions will have been met for the mass production and establishment of powerful automated information systems on an industrial scale. From that time on, the entire knowledge industry complex will have access to these new facilities and will use them extensively. As a result it will undergo rapid, radical change and its future will then seem very different from what we believe it to be in the present.

In terms of numbers and processing capacity, the electronic information systems created to meet the varied needs of the knowledge industry will in 1985–87 be almost a hundred times those of today. However, today's systems will be hardly comparable with those of the future. In any case, multiplication by a factor of 50 by comparison with 1970 seems a reasonable assumption . . .

(1) In all the case studies, growth follows a geometric progression, the curve being exponential.

(2) The growth rates observed however varied considerably, the lowest one 3·5% yearly, the highest 14·4%.

(3) The lowest growth rates are of the number of scientific periodicals published, covering a 300-year period, and the number of specialised bibliographical periodicals involved in indexing and abstracting over a 140-year period. In the case of scientific journals, the annual growth rate has been 3·5%, 3·7% or 3·9% depending whether the number published in 1972 is taken as 30 000, 50 000 or 100 000. The growth rate for indexing and abstracting organisations has been 5·5% a year. In 1972, there were 1800 such services in science.

(4) A recent series reporting the number of articles by engineers in civil engineering journals (from 3000 pages of technical articles in three specialized periodicals in 1946 to 30 000 pages in 42 specialized periodicals in 1966) shows growth rates of 12·3% a year.

(5) The growth rate in the number of international scientific and technical congresses increased almost fourfold in 20 years, rising from 1000 in 1950 to over 3500 in 1968 . . .[13]

The US National Academy of Sciences estimated that in the early 1970s about 2 million scientific writings were published every year – i.e. 8000 for each of 250 working days – and that the figure would quadruple by 1985. Anderla noted, however, that this should be 'considered as a conservative forecast'; that a continuation of the 1967–71 trends would produce 'a total of some 13 to 14 million per year, i.e. equivalent to the stock accumulated since the origins of science until the present day. A projection midway between these figures would seem fairly reasonable.' The mathematician Stanislaw Ulam found that in the 1970s about 200 000 new theorems were published each year in mathematical journals and decided that this was 'something to worry about . . . If the number of theorems is larger than one can possibly survey, who can be trusted to judge what is "important"?'

During manned space flights, data was transmitted to earth at the rate of 52 kilobits per second – the equivalent of an *Encyclopedia Britannica* every 79 minutes. The technical capacity to process information has grown exponentially, but human capacity to select, interpret and apply material cannot expand at a comparable rate.[14]

This poses many problems: vast increases in the production of information of all kinds – newspapers, television, magazines, pop music, computer print-outs, promotional and advertising campaigns, forms to be filled in – may make it harder for people to feel that they can 'cope' or keep up. It is no coincidence that there is a vast increase in stress-induced diseases – with a high incidence of mental breakdown, suicide, alcoholism and obesity. In his spirited polemic *The Costs of Economic Growth* (Penguin, 1969), Professor E. J. Mishan points to the growing speed of the obsolescence of personal skills, which is a corollary of the growing speed of technological advance.

> Every one of us . . . lives closer to the brink of obsolescence. Each one of us that is adult and qualified feels menaced in some degree by the push of new developments which establish themselves only by discarding the methods and techniques and theories that he has learnt to master . . . Inasmuch as experience counts for less and knowledge, up to date knowledge, far more in a world of recurring obsolescence, the status of older men falls relative to that of younger men. And within the family the same force is at work . . . The rapidity of change in social conventions and moral attitudes, associated with technological transformations in the mode of living, renders a person's experience of the world a generation ago largely irrelevant to the problems of the day.

Trade or professional training is likely to have strictly limited relevance. Lord Ashby has suggested that university degrees and diplomas should be, like passports, renewable after five years: they can no longer be regarded as a lifetime qualification. PhDs in engineering, for example, have a 'half life' of less than ten years. People may do their best professional work shortly after qualifying and begin a long period of decay as they enter their thirties, by which time they are gently shunted aside into less exacting tasks (e.g. administration) to make way for the gifted younger people coming on.

Alvin Toffler used the term 'future shock' to illustrate sensory and information bombardment of the human brain. In his article 'The Magical Number Seven, Plus or Minus Two: Some Limits on our Capacity for Processing Information', George A. Miller argued, after examining a mass of evidence, that there is a 'span of absolute judgment' for identifying 'the magnitude of a unidimensional stimulus variable' – such as ranking pitch, loudness, colour, tones and taste.[15] For most people this span is seven: the more gifted may have a span of nine, the less gifted of five. (Where material is more complex and interrelated the number of items in the span increases rapidly, but there are still upper limits: the computer is not troubled by such limitations.) People – like Pavlov's celebrated dogs – may feel unable to make rational choices from the vast array of information thrust at them, and fall back on mass responses (i.e. making the same choices as their neighbours).

We have come to accept the universality of television almost without question, and to assume that it has meant enrichment of life: for many, especially the aged or house-bound, this is certainly true. With television as an integral part of life, however, it becomes almost irrelevant to ask whether we have too much, or whether the quality is too low. Television rarely examines the possibility that there may be better ways of using one's time (the 'Life. Be In It' campaign may be an exception). If, fifty years ago, we had been able to hover over a city in a balloon and were told 'In this city, 80 per cent of the inhabitants are now reading but they read only four books', we could be certain that the city was either part of a theocracy or that it was dictatorship. What can be said of a society in which 80 per cent of its inhabitants freely chose to watch four highly repetitive channels? Has the universal presence of television made the balloon story irrelevant? In 1979 an Australian Broadcasting Tribunal survey found that 63 per cent of those polled thought that television news gives the best coverage, 22 per cent newspapers, 15 per cent radio. Sixty-one per cent thought that television news was mostly accurate, and 72 per cent thought it was detailed enough. This is getting very close to what Herbert Marcuse called 'one-dimensional society', where a range of choice may be so limited as to be virtually no choice at all (e.g. Jerry Ford or Jimmy Carter in the 1976 Presidential election).

Karl Marx coined the word 'alienation' to describe a situation where people felt remote or estranged from their environment, having a sense that 'I don't understand what's going on any more. I feel lost.' Emile Durkheim used the word 'anomie' to describe a lack of direction or purpose, a condition of aimlessness or rootlessness. Both conditions are products of the speed and variety of change, which disturb the relationship of people with their environment. Changes which might have occurred over a generation now occur within a few years – the pace is forced. Paradoxically, as the globe shrinks through instant electronic communication and supersonic travel there is a growing sense of estrangement and isolation between ordinary people, especially in cities. Many young people are unable to form a coherent view of their role or potential in an 'other-related' context. Many find that the education process is meaningless – they are trained for life in an industry which may have changed, or even disappeared, by the time they have spent a few years at work. Others are involved in boring, repetitive and meaningless work. A significant proportion can find no work at all.

INFORMATION – WHO CONTROLS IT?

Unless policies are adopted which encourage a two-way flow of information, the relative position of individuals *vis-à-vis* institutions may deteriorate rapidly. To laymen, the basic problem may be working out which of a bewildering array of keys should be used to open the right door. For Marxist analysis, control of data banks is as appropriate an illustration of class struggle as is control of a conventional bank, which is increasingly becoming a repository of information rather than bullion. Does control of, and access to, information resources necessarily have to be the preserve of the few? There are significant barriers to egalitarian access to information in Australia.

We have only experienced the first stages of the 'information revolution'. The next stage will see a further integration of computerization and telecommunications – 'telematics' as Nora and Minc call it – which will provide instant access to information stored anywhere in the world to anybody who can operate a telephone. Dr Yoneji Masuda has stressed the potential of computer-based information systems, if properly used, to strengthen the position of individuals. Whereas he sees mass communication information as 'passive ... uniform ... used only once in a one-way traffic', information produced by the computer is a two-way exchange: 'Moreover, feedback is possible and the information is tailor made to the specific needs of the user. The most conspicuous feature is the information user becomes the constituent in information selection.' In an essay 'Future Perspectives for Information Utility', Masuda endorsed Kenneth Boulding's point about how knowledge defies the second law of thermodynamics:

> First of all, information, unlike goods, has four inherent properties, that have made self-multiplication possible. Information is (1) inconsumable, (2) untransferable, (3) indivisible, (4) accumulative.
>
> (1) Inconsumable – Goods disappear through use. Information does not disappear but remains unchanged however much it is used.[16]
>
> (2) Untransferable – When a good is transferred from A to B, it is moved completely from A to B, but when information is transferred from A to B, the original information remains at A.
>
> (3) Indivisible – Goods used as materials (electricity, water, etc.) can be divided and used, but information can only be used when it constitutes 'a set'.
>
> (4) Accumulative – The only way to accumulate goods is to not use them. Information, however, because it cannot be consumed or transferred, can be accumulated while it is to be used again and again. Information of a higher quality is produced by adding new information to the information that has been accumulated previously.
>
> Although information has always had the property of self-multiplication, computer-communication technology has rapidly increased the speed and quality of self-multiplication because the technology itself has added four more properties to information: (1) concentration, (2) dispersion, (3) circulation, and (4) feedback.[17]

James Martin and Adrian Norman write that

> telecommunication links will bring the capabilities of the computers to the millions of locations where they can be used, and computers in return will control the immense switching centres and help divide the enormous capacity of the new linkages into usable channels . . .
>
> The telephone lines and probably other, new cables, will connect our homes and offices to the computer. One will be able to dial machines using particular programmes or data just as today one can dial a friend . . . This will change our working patterns just as thoroughly as the first Industrial Revolution changed them.[18]

Who should control international data networks? Should it be a monopoly in the hands of telephone corporations, mostly state enterprises such as Telecom Australia or the British Post Office, or controlled by private enterprise of which the largest, richest and best-known example is American Telephone and Telegraph (AT&T)? Alternatively, should a rival telecommunications network be set up to compete with the telephone systems, able to use a variety of the latest techniques – communications satellites, laser beams, optical fibres (in which signals are transmitted by light waves), high-frequency coaxial cables, ultra-high-frequency (UHF) mobile networks, or wave-guides (pipes containing thin glass fibres)? The question of control of, and access to, information should become one of the major political issues of the 1980s. If it does not it will be a national tragedy: if we pass through 1984 without even discussing the issue, it may mean that the battle for control has – unnoticed – already been won by oligopolists and centralizers.

There are two major options – and a variety of smaller ones in between. First, and most likely, information networks will be centralized, oligopolist and limited – essentially a means of preserving existing power structures, and controlled by the same people who own newspapers, television networks and radio stations and who see the computer (with some justification) as an instrument of counter-revolution. Second, and much preferable, information networks will be regarded as a public utility, open to all who can pay an appropriate low-cost fee for data. There is heartening evidence that existing networks can be used to enable small firms to provide highly specialized services in computer programming and consulting (e.g. in regional or provincial areas, assisting development of medium-sized labour-absorbing projects). Personal computers, public international data networks, computer service bureaux and the electronic media can give the individual access to information for the same price available to large organizations.[19] Computer organizations in the future will provide access to product data bases for use by consumers through terminals in their homes.

The logical extension of equal access to information would be to create far more open socio-economic structures – where a consumer organization could have the same access to information as a tyre manufacturer, trades unions as employers, an opposition party as a government, backbenchers as parliamentary ministers, pacifists as the military, media users as media proprietors, or a general practitioner in a poor country as a specialist in a rich one. The existing information infrastructure is now far too strongly established to be dismantled, but it can be made democratic and passive consumers can become active users of and contributors to the knowledge supply. The sheer volume of available data need not perplex us, providing that we can set our own pace in making use of it. Using STD (subscriber trunk dialling) or ISTD (international STD) is hardly more complicated than dialling locally – although the technology is far more complex and the range of choices enormously greater. A library with ten million books is no more intimidating than one with only a million.

In 1972 JACUDI published 'The Plan for a National Information Society – a National Goal Towards the Year 2000', based on a work by Kenichi Koyama. The plan recommended a framework for a computerized socio-economic restructuring of Japan at an estimated cost of $US67 000 million: it provided for the rationalization of marketing, administration, medicine, housing needs, traffic regulation, pollution controls and information resources. The Japanese government rejected the plan in 1974 as 'too expensive and too extreme'. JACUDI then proposed options for four communications networks – industrial, social (maximizing access to data on health, education and housing), leisure (tourism, recreation, sports, craft and developing new life-styles) and information (publication, radio, calculation, research and data processing). The Nora Report described the JACUDI plan as

> intensely interventionist and innovative. It does not nationalise any sources of supply, but it does nationalise a growing portion of demand . . . It is based on . . . an absence of distrust with regard to automation. Thus this project is based on a type of relationship between the state and industry, a social consensus, a national determination and an absence of individualisation that does not make it suitable for generalisation . . . The defects of this project are its enormousness and its one-sidedness.[20]

In 'Future Perspectives for Information Utility', Yoneji Masuda proposed setting up more modest public-information processing and service facilities combining computer and communications networks, so that anyone, anywhere and at any time can easily, quickly and inexpensively obtain the information he needs. Masuda recommended that such a utility comprise the following elements:

1 The central facilities are equipped with a large scale computer capable of simultaneous parallel processing, and to which are attached large capacity memory devices, a large number of program packages and extensive data bases. The facilities must be able to carry out information processing and service for a large number of users simultaneously.

2 Because these information process and service facilities are provided for the use of the general public, the computers of the central facilities are connected by means of a communications circuit directly to terminals in the businesses, schools, and homes that utilize the facilities.

3 At any time the user can call the local centre of the information utility to process data himself or to have the centre process and service the necessary data for him.

4 The rate charged for using the centre must be low enough that the general public can use the centre readily for day-to-day goals.

In the future information society the time will come when the general public will be able to use freely any of the information utilities in hundreds of thousands of places by simply using the terminals in the places where they live. These information utilities are not existing in actuality now, but we can see their beginnings around us . . .[21]

He concluded that 'autonomous management by citizens of the information utility is an absolutely indispensable prerequisite for the ideal information society', and that in order to achieve this 'we must concentrate every effort on resolutely ridding the information utility of dictatorial domination by special classes of power'.

Nora and Minc recognize that telematics can be centralized or decentralized. They have no doubt that miniaturization should be used to convert information-processing from 'an elitist technique' to a 'mass activity'.

Today, any consumer of electricity can instantly obtain the electric power he needs without worrying about where it comes from or how much it costs. There is every reason to believe that the same will be true in the future of 'telematics'. Once the initial connections are made, the network will spread by osmosis . . .

Traditional data processing was hierarchical, isolated and centralised. The technical constraints were prejudicial in terms of the mode of organisation which it imposed, because the presence of computers relieved the natural unwieldiness of enterprises and administrations . . . From now on, data processing can be deconcentrated, decentralised, or autonomous: it is a matter of choice. . . .

Information technology has today become an almost completely flexible tool. Its organisation can spread without encountering a major obstacle through all the configurations of power. It will disrupt the rules and conditions governing competition among numerous economic agents; it will confirm or annul the relative importance of the centre and the periphery in most organisations. But this diffused penetration will involve deep changes in essential functions (medicine, education, law, Social Security, working conditions), and by increasing openness will bring into question the freedoms and inherited privileges of the dark areas of society.[22]

Even before the Nora Report, France had begun a technological revolution in telecommunications with the world's first electronic time division multiplexing (tdm) switching centre in 1970, and $30 000 million has been invested in upgrading and transforming existing networks. Transmission is being converted from analogue to digital, and the use of communication satellites makes international transmission and reception of data in oral, visual and written form reliable and cheap. In the 1980s, France's telematic programme will offer electronic telephone books, home shopping (push-button choice of goods shown on a VDU), electronic mail, an information bank, phone banking, facsimile newspapers, home alarm systems, instant self-service travel bookings, all at low cost.

The telematic revolution will either threaten Australia's media oligopoly or consolidate it, depending on political responses to technological change. Australia has a heavier concentration of media ownership than any Western country (even Ireland): three major groupings dominate metropolitan daily newspapers – Herald & Weekly Times Ltd (52·7 per cent of circulation), News Ltd (Rupert Murdoch, 24·7 per cent), and John Fairfax Ltd (22·6 per cent). Together with Australian Consolidated Press (Packer family), the same interests control radio stations, television channels, magazines and suburban newspapers. Would they control Australia's proposed communications satellite and/or national or international information services? If matters are allowed to drift, the present pattern of media control will be extended and intensified. (In the United States and Great Britain, press and television ownership are separated and elements of the media tend to be fierce rivals rather than complaisant allies. There is a far greater tradition of radical dissent and citizen participation in both countries. Where such traditions have evolved in Australia, they have frequently been 'legitimated' because of overseas activity rather than a spontaneous domestic reaction: another example of the 'cultural cringe'.) The principal stated reasons for an Australian communications satellite are to provide telephone services for 40 000 people in the outback excluded from the terrestrial network, improve country television reception, assist with emergencies, and improve delivery of health and education services. Almost as an afterthought, the Minister for Post and Telecommunications (Tony Staley) added that another reason was 'significantly increased capacity for distribution of bulk speed data communications'.[23]

The threatening potential implicit in the new information technology must also be considered. In the US there was a major debate in 1967 about the implications of setting up a national data centre, with controls of centralized information on personal credit, health, educational, banking and insurance records. In Australia we have not had the debate, but the technology and relevant computer records for a national data centre are already in operation. Joseph Weizenbaum has pointed to US research on computerized speech recognition: this would make it possible to monitor millions of telephone conversations in which the use of key words – perhaps a politician's name, a reference to sexual deviance, advocacy of militant industrial action, discussion of defence matters or relations with foreign powers – could trigger off a recording and/or transcript for the edification of those in authority. Such techniques bring us close to the world of Big Brother, the Ministry of Truth and 1984. Martin and Norman observe that

> cyberveillance techniques could be used to keep watch on telegrams, bank-account postings, credit transactions, licences, share purchases, airline travel, hotel bookings, car rentals and so on. Some official, somewhere, invariably has a good reason for using such information . . . The only limit to cyberveillance is cost, and that is dropping all the time.

They point out the value of the German census to Hitler as a means of extending totalitarian power: a computer network in the hands of a Hitler or Stalin would multiply this infinitely.[24]

> The superb efficiency of future automation will facilitate the exercise of intolerance – either overtly by governments or as a by-product of other well-meaning social forces. The automation of intolerance is an alarming prospect. We suspect that as the power of the machines increases and the cost of data banks decreases, we shall, whether we like it or not, lose some degree of our present privacy concerning personal information. The forces bringing the data banks into being are greater than the forces seeking to prevent them. If this is so, and if the means to automate intolerance do exist then we believe that the computerized society will have to become a far more tolerant society than most societies of the past if true freedom is to survive.[25]

The way out of the flybottle

The Australian constitution empowers the Commonwealth Parliament to make laws about 'postal, telegraphic, telephonic, and other like services'.[26] The High Court has interpreted this to cover radio and television, and presumably would extend it to the satellite as well. However, there is no power to regulate newspaper ownership *per se*. Telematics, assisted by communications satellites, can expand or cripple human capacity and understanding. Monopolist or oligopolist control of information is as unthinkable as entrepreneurial control of air, water or sunlight for profit.

The Australian communications satellite must be a common carrier, open to all potential users and not used to increase the power of the strong against the weak, or as an instrument of manipulative control over passive but receptive consumers. Its ownership should be vested in an Australian Information Utility. Corporate interests should not necessarily be excluded from sharing control of such a utility, but they must not dominate it. Government, trade unions and other special-interest groups such as consumers, environmentalists, women and ethnic minorities must be part of the utility management and able to share in the collection, dispersal and feedback of information.

There must also be a complementary (and to some extent conflicting) legislation to provide freedom of information and guarantee privacy. The former may strengthen the weak against the strong, but without guarantees of privacy and limitations on the type of personal data collected and stored, the computer controllers will gain information about individual citizens to a degree which is unprecedented and totally inconsistent with democratic theory or practice. The political programme set out in Chapter 11 includes, in broad outline, further recommendations which would ensure that modern technology is used to democratize information flow and that provisions are made to safeguard individual privacy.

Where is the wisdom we have lost in knowledge?
Where is the knowledge we have lost in information?

T. S. Eliot, Chorus from *The Rock*

9 Work in an Age of Automata

For even when we were with you, this we commanded you, that if any would not work, neither should he eat.

St Paul, II Thessalonians iii, 10

Historical attitudes to work

In the Bible, God imposed work as a punishment on Adam and Eve for their disobedience in eating fruit from the tree of knowledge. The loss of innocence involved in this 'fall' led to God's curse on Adam: 'In the sweat of thy face shalt thou eat bread, till thou return unto the ground: for out of it wast thou taken: for dust thou *art*, and unto dust shalt thou return.'[1] God expelled him from the Garden of Eden, and commanded that he 'till the ground from whence he was taken'. In the Greek creation myth the beautiful Pandora, the first woman created by Zeus, was given a lidded vase containing all the ills afflicting humanity and was told not to open it. She disobeyed, and the evils flew out. One of them was *work*, in classical Greek *ponos* (ποηος)', the root from which the words 'pain' and 'punishment' are derived. In the 'golden age' of Periclean Athens (460–429 BC), a relatively small leisured class enjoyed a rich and varied cultural and recreational life, supported by members of a large servile class – technically slaves, but more like servants – who performed all routine work. Sir Alfred Zimmern denied that the Greeks of that period regarded manual labour as degrading:

> In truth they honoured manual work far more than we do . . . But they insisted, rather from instinct than policy, on the duty of moderation, and objected, as artists do, against doing any more work than they needed when the joy had gone out of it. Above all they objected to all monotonous activity, to occupations which involved sitting for long periods in cramped and unhealthy postures . . .[2]

While the Greeks accepted the maxim of 'art for art's sake', they did not advocate work for work's sake and saw no particular moral value in it. Aristotle (384–322 BC) considered work to be an unnecessary interference with a citizen's higher duties – the pursuit of virtue and truth, cultivation of the arts, and participation in public affairs. In *Poetics*, Aristotle considered the possibility of abolishing slavery but concluded:

> There is only one condition in which we can imagine managers not needing subordinates, and masters not needing slaves. This condition would be that each [inanimate] instrument could do its own work, at the word of command or by intelligent anticipation, like the statues of Daedalus or the tripod made by Hephaestus, of which Homer relates that 'of their own motion they entered the conclave of Gods on Olympus', as if a shuttle should weave of itself, and a plectrum should do its own harp-playing.[3]

Aristotle's contemporaries would have thought his examples far-fetched, but the condition he wrote of – instruments which could do their own work – has now been met. What effect will it have on our attitude to work?

Western civilization was decisively shaped by Greek culture – but not in attitudes to work and economics, which are derived far more from the Judaeo-Christian tradition.[4] Jesus never referred to Adam's curse, but it is implicit in St Paul's second letter to the Thessalonians where he preaches against idleness and disunity within the tiny Christian community.[5] The rise of Christianity coincided with the contraction of slavery. P. D. Anthony wrote:

> An ideology of work is redundant when the labour force can be conscripted and coerced at will. In conditions of a freer labour market an ideology has to be developed in order to recruit labour and then in order to motivate it by persuading it that its tasks are necessary or noble. In conditions of a free market and a chronic shortage of labour, the manufacture and communication of an ideology of work becomes a central preoccupation of society.[6]

The Stoics, Virgil, Cicero and the early Christians questioned the moral basis of slavery and promoted the idea of self-sufficiency and the dignity of labour. In imperial Rome, the working day was seven hours in summer, six in winter. There were ninety-three public holidays with state-funded spectacles under Augustus, and forty-three *feriae publicae* (essentially religious in origin), leaving 227 working days each year. Marcus Aurelius established 230 days as the standard business year. The Circus Maximus had 250 000 seats, enough to accommodate a quarter of Rome's total population, which confirms the importance of leisure pursuits ('bread and circuses') at that time. Under Claudius there were 159 workless days, while the Calendar of Philocalus (AD 354) records 200 holidays.[7]

Professor Sol Encel has pointed out:

> St Benedict, in the 6th century, praised work as a means towards achieving grace. 'Work', he said, 'and do not despair.' One of his monastic rules declares that idleness is the enemy of the soul. The Benedictines reinforced their

devotion for work through a regular daily schedule in which prayer and work alternated, governed originally by the ringing of bells and later through the use of the clock.[8]

To St Thomas Aquinas, society was based on a harmonious relationship between all classes, and the good life was essentially co-operative. High interest and excessive profit were disreputable; work could be noble, but the rich should not exploit the poor. A 'just' price for commodities should be set so as to be fair to both producer and consumer. Competition was not a good thing because it encouraged greed, luxury and harsh dealing. Emmanuel Le Roy Ladurie, in *Montaillou*, his famous study of life in a medieval village in the year 1308, found that the 'people tended to shorten the working day into a half day' and 'were fond of taking a nap . . . of taking it easy'.[9] In *Utopia* (1516) Sir Thomas More described work as a universal obligation but regarded a six-hour day as sufficient to satisfy Utopia's needs. Tommaso Campanella's romance *The City of the Sun* (1623) predicted a four-hour day.

The work ethic which still dominates the minds of political leaders and much of society in advanced economies is often called the 'Christian', 'Protestant' or 'Puritan' work ethic – although it is characteristic of Chinese, Japanese and Jews just as much as Western Europeans or North Americans. The German sociologist Max Weber, in *The Protestant Ethic and the Spirit of Capitalism* (1905), made a fundamental criticism of Marx's thesis that religion is essentially a form of cultural expression, determined by the economic structure of society. Weber argued that there was a more complex interrelationship between economic relations and religious belief – and that the Reformation of Luther and Calvin represented the explosion of a spirit of individualism, self-advancement and competition which encouraged the growth of a 'philosophy of avarice' in which 'the increase in capital is assumed as an end in itself, in which economic acquisition is no longer subordinated to man as the means for the satisfaction of his material needs'. This led to a 'hard frugality in which some participated and came to the top, because they did not wish to consume but to earn'.[10] Luther, to some extent, and Calvin, to the point of obsession, advanced the idea of a 'vocation' to which God calls men – and argued that frugal living and sober economic advance might well be visible signs of God's grace in choosing them to be among the elect. Luther said that grace could be lost and regained. Calvin believed that 'election' was predetermined and thus unchangeable – nevertheless his followers in Geneva and elsewhere felt convinced that capital gains must surely be a reflection of divine satisfaction and a sign of redemption. As Weber wrote, 'the puritan outlook favoured the development of a rational bourgeois economic life' – and it marked an end to medieval economics in northern Europe. In *Religion and the Rise of Capitalism* (1948), R. H. Tawney wrote:

When the age of the Reformation begins, economics is still a branch of ethics, and ethics of theology; all human activities are treated as falling within a single scheme, whose character is determined by the spiritual destiny of mankind; the appeal of theorists is to natural law, not to utility; the legitimacy of economic transactions is tried by reference, less to the movements of markets, than to moral standards derived from the traditional teaching of the Christian Church.

The links between Protestantism and capitalism can be exaggerated – the great Italian banking families of the fifteenth century, such as the Medici, and the Fuggers of Augsburg, were Catholics – but they should not be understated either. Dr Octavio Paz has compared the impact of Hispanic Catholicism and English Protestantism in America with their

contrasting attitudes towards work, festivity, the body and death. For the society of New Spain [i.e. Mexico and South America] work did not redeem and had no value in itself. Manual work was servile. The superior man neither worked nor traded. He made war, he commanded, he legislated. He also thought, contemplated, wooed, loved and enjoyed himself. Work was good because it produced wealth, but wealth was good because it was intended to be spent . . . Work is the precursor of the fiesta. The year revolves on the double of axis of work and festival, saving and spending . . . The United States has not really known the art of the festival . . . This is natural. A society that so energetically affirmed the redemptive value of work could not help chastising as depraved the cult of the festival and the passion for spending . . . Work is purification, which is also a separation . . . Capitalism exalts the activities and behaviour patterns traditionally called virile: aggressiveness, the spirit of competition and emulation, combativeness.[11]

Work and the division of labour

Neolithic gold, bronze and ironwork of surpassing quality – much of it discovered after World War II – testifies to the creative skills of craftsmen long before the development of writing. Cypriot Bronze Age pots are among the commonest of archaeological relics, and yet each one of the many thousands in museums throughout the world bears marks which identify its maker – a personal 'trade mark', or a thumb identation. We may assume that the potter took pride in his work and that creative satisfaction was central to his life, not peripheral.

The rise of capitalism and its extension throughout the Atlantic world, followed by increasing urbanization, the onset of the First Industrial Revolution and the rise of socialism, were inextricably linked with increasing division of labour which changed the whole nature of work. Until the Industrial Revolution, craftwork – whether on farms or in towns – was absolutely central to human experience. The potter, the carpenter, the stonemason and the metal-worker were both tradesmen and creative artists. Although working within a recognized genre, such as making a water-jug, the potter was in charge of the whole production process – from choosing raw materials, working out his own design, shaping the jug, decorating and firing it, to rejecting poor-quality goods and selling the remainder. The tragedy of the Industrial Revolution and its specialization has been that 'manufacturing' has become 'machinofacturing', and that for nearly two hundred years workers have been physically prevented from taking their products through from conception to execution. Few process workers in motor-manufacturing plants are able to identify themselves with a particular vehicle and say with pride, 'I put the left front hubcap on that Toyota'.[12]

The destruction of craftwork has led to specialization, but at the cost of a general de-skilling. By making creativity remote from the work experience, it has made aesthetic considerations seem to be the preserve of an educated elite. As individuals lose control of the production process they inevitably become integrated in a larger, more complex economic system – in which their own role becomes fragmented and complementary. The concentration of economic power, creating a hierarchy of command, is a corollary of large-scale industry and commercial or market-based agriculture. Harry Braverman wrote of American farms before 1810 that, in addition to producing and processing crops and livestock,

> the farmer and his wife and their children divided among them such tasks as making brooms, mattresses, and soap, carpentry and small smith work, tanning, brewing and distilling, harness making, churning and cheese making, pressing and boiling sorghum for molasses, cutting posts and splitting rails for fencing, baking, preserving, and sometimes even spinning and weaving.[13]

But specialized farm production led to the narrowing and loss of traditional skills, the 'transformation of society into a giant market for labour and goods', and a mental attitude which turned ' "home made" into a derogation and "factory made" . . . into a boast'. He estimated that in the early 1800s perhaps 80 per cent of the US labour force was self-employed and owned the means of production, but that this figure fell to 33 per cent in 1870, 20 per cent by 1940 and 10 per cent by 1970.[14] Since he defined the 'working class' as persons who were not owners of capital and were not self-employed, he concluded that 'some two-thirds to three-fourths of the total [US population] appears readily to conform to the dispossessed condition of a proletariat', and that the 'petty bourgeois of pre-monopoly capitalism now corresponds increasingly to the formal definition of a working class'.

Braverman's logic is impeccable, but his description would be rejected indignantly by the majority of Americans who do not consider themselves 'working-class' and who point to the high proportion of home ownership, and rising consumption levels. A man who owned a block of land, a spade and a cow in 1810 might have been self-employed and technically a capitalist – but he may have been near destitution for much of the year, needing to hire himself out as a day-labourer to make ends meet. An engineer or bank manager of the 1980s – employees in a hierarchical system – may have accumulated capital assets but not capital for investment, and will often look down on the blue-collar worker, identifying himself with the employer rather than with fellow employees.

SMITH'S PHILOSOPHY OF WORK

Adam Smith was not an advocate of work for work's sake – he took a position much closer to that of the Greeks. He saw work as a regrettable necessity and regarded manufacturing as debasing. Smith argued that high wages would lead to 'the greatest public prosperity' and encourage population growth, and that labour should be 'of moderate duration'. He argued that education be made generally available in England (as it had been in some areas of Scotland for generations) and thought that too much division of labour would reduce the worker 'to a remarkable degree of stupidity in which he is not able to exercise his civic duties'. He urged *laissez-faire* because he thought that, in the long term, self-interest on the part of the capitalists would make them recognize that a prosperous, skilled and free labour force would contribute more to economic growth than a demoralized, poverty-stricken one. But in the long term, as Keynes said, we are all dead – and Smith's disciples ignored his irony and concern for non-economic ends, using their power to degrade working conditions, lengthen hours, and displace male adult workers by women and children.

MARX'S PHILOSOPHY OF WORK

Marxists and capitalists share more philosophical positions than is generally recognized:

1 the unquestioning assumption that work is necessarily a good thing in itself;

2 acceptance of technological determinism;
3 the concentration of economic power;
4 the creation of a hierarchy of command;
5 the priority of economic ends.

The young Karl Marx, repelled by the economic reality he saw around him, denounced the concept of 'alienated labour':

> According to the laws of political economy the alienation of the worker in his object is expressed as follows: the more the worker produces the less he has to consume, the more values he creates the more valueless and worthless he becomes, the more formed the product the more deformed the worker, the more civilised the product, the more barbaric the worker, the more powerful the work the more powerless becomes the worker, the more cultured the work the more philistine the worker becomes and more of a slave to nature . . .
>
> The product of labour is externalisation . . . labour is exterior to the worker, that is, it does not belong to his essence . . . Thus the worker only feels a stranger. He is at home when he is not working and when he works he is not at home.[15]

However, Marx came to see work as being intrinsically good when men had the chance to grapple with and solve problems as a form of self-realization, and he took up this theme in the *Grundrisse*, Notebook VI (1857–58).[16] In Volume III of *Capital*, Marx argued:

> The actual wealth of society, and the possibility of constantly expanding its reproduction process, therefore do not depend upon the duration of surplus-labour, but upon its productivity and the more or less copious conditions of production under which it is performed. In fact, the realm of freedom actually begins only where labour which is determined by necessity and mundane considerations ceases; thus in the very nature of things it lies beyond the sphere of actual material production. Just as the savage must wrestle with Nature to satisfy his wants, to maintain and reproduce life, so must civilised man, and he must do so in all social formations and under all possible modes of production. With his development this realm of physical necessity expands as a result of his wants; but, at the same time, the forces of production which satisfy these wants also increase. Freedom in this field can only consist in socialised man, the associated producers, rationally regulating their interchange with Nature, bringing it under their common control, instead of being ruled by it as by the blind forces of Nature; and achieving this with the least expenditure of energy and under conditions most favourable to, and worthy of, their human nature. But it nonetheless still remains a realm of necessity. Beyond it begins that development of human energy which is an end in itself, the true realm of freedom, which, however, can blossom forth only with this realm of necessity as its basis. The shortening of the working-day is its basic prerequisite.[17]

In the same volume he predicted the rise of a white-collared managerial and professional class entirely divorced from the ownership of capital: 'An orchestra conductor need not own the instruments of his orchestra . . . The capitalist's work does not originate in the purely capitalist process of production . . . [but] from the social form of the labour process.'[18]

In *Grundrisse*, Notebook VI, Marx referred to the 'last . . . metamorphosis of labour . . . the *machine* or rather an *automatic system of machinery*':

> This is set in motion by an automaton, a motive force that moves of its own accord. The automaton consists of a number of mechanical and intellectual organs, so that the workers themselves can be no more than the conscious limbs of the automaton. In the machine, and still more in machinery as an automatic system, the means of labour is transformed as regards its use value, i.e. as regards its material existence, into an existence suitable for fixed capital and capital in general; and the form in which it was assimilated as a direct means of labour into the production process of capital is transformed into one imposed by capital itself and in accordance with it. In no respect is the machine the means of labour of the individual worker. Its distinctive character is not at all, as with the means of labour, that of transmitting the activity of the worker to its object; rather this activity is so arranged that it now only transmits and supervises and protects from damage the work of the machine and its action on the raw material.
>
> With the tool it was quite the contrary. The worker animated it with his own skill and activity; his manipulation of it depended on his dexterity. The machine, which possesses skill and force in the worker's place, is itself the virtuoso, with a spirit of its own in the mechanical laws that take effect in it; and, just as the worker consumes food, so the machine consumes coal, oil, etc. (instrumental material), for its own constant self-propulsion. The worker's activity, limited to a mere abstraction, is determined and regulated on all sides by the movement of the machinery, not the other way round. The knowledge that obliges the inanimate parts of the machine, through their construction, to work appropriately as an automaton, does not exist in the consciousness of the worker, but acts upon him through the machine as an alien force, as the power of the machine itself.
>
> Invention then becomes a branch of business, and the application of science to immediate production aims at determining the inventions at the same time as it solicits them. But this is not the way in which machinery in general came into being, still less the way that it progresses in detail. This way is a process of analysis – by subdivisions of labour which transforms the worker's operations more and more into mechanical operations, so that, at a certain point, the mechanism can step into his place.
>
> Thus we can see directly here how a particular means of labour is transferred from the worker to capital in the form of the machine and his own labour power devalued as a result of this transposition. Hence we have the struggle of the worker against machinery. What used to be the activity of the living worker has become that of the machine.
>
> Thus the appropriation of his labour by capital is bluntly and brutally presented to the worker: capital assimilates living labour into itself 'as though love possessed its body'.[19]

Marx grasped earlier than anyone else that the division of labour, taken to the ultimate, would dig a grave for unskilled and semi-skilled workers because the highly specific, repetitive tasks created by the atomization of work are capable of being automated with great precision and low cost. But he also recognized the revolutionary transformation that could occur because of increasing leisure time.

From the standpoint of the immediate production process free time may be considered as production of fixed capital; this fixed capital being man himself. It is also self-evident that immediate labour time cannot remain in its abstract contradiction to free time – as in the bourgeois economy. Work cannot become a game, as Fourier would like it to be; his great merit was that he declared that the ultimate object must be to raise to a higher level not distribution but the mode of production. Free time – which includes leisure time as well as time for higher activities – naturally transforms anyone who enjoys it into a different person, and it is this different person who then enters the direct process of production.[20]

The most vehement attacks on Marx come from those who have never read him. Although often infuriating, perverse and wrong, he provided an incomparable historical analysis of the industrial era which is worth consulting as we pass through post-industrialism into a post-service era.[21]

Why work?

Why do we work? Is work an inevitable part of human experience? How could society cope with a substantial permanent reduction of work? The far Left and the far Right share a fundamentalist commitment to the traditional work ethic, in either its Marxist or Christian form, but they apply the principle selectively. Marxist dogmatists insist that work should be compulsory under communism but not under capitalism, while capitalists see it as a good thing under private enterprise and deplorable under socialism.[22] Both positions assert that work is a good thing in itself rather than a matter of free choice, and we may infer from this that a fifty-hour week is to be considered as better than one of forty, and forty as better than thirty. Many trade unionists fear that a reduction from forty hours' work (plus overtime) to thirty hours will inevitably lead to a 25 per cent income drop – and given a choice between longer hours and reduced purchasing power, will opt for the first.

It is hard to identify any beneficiaries of the work ethic. For those who feel that loss of employment is a form of social death the principle of compulsory work is a torment, forcing them in search of jobs that no longer exist and reinforcing a prevailing conviction that the unemployed are worthless and must be penalized. There is an almost sadistic view that if whips and scorpions are applied to the unemployed, they will spring out of a wicked lethargy and take up employment: therefore, it is argued, a punitive approach to unemployment is in the long-term interests of the unemployed themselves.[23] The majority of the unemployed are unskilled and poorly educated, in an economy in which unskilled, routine and repetitive jobs are declining rapidly. Forcing them into short-term meaningless boondoogling jobs may be economically pointless but may be justifiable in social and psychological terms.

Bertrand Russell made a fundamental attack on the cult of work in his essay 'In Praise of Idleness':

> I think that there is far too much work done in the world, that immense harm is caused by the belief that work is virtuous, and that what needs to be preached in modern industrial countries is quite different from what always has been preached . . .
>
> The morality of work is the morality of slaves, and the modern world has no need of slavery . . .
>
> [World War I] showed conclusively that, by the scientific organisation of production, it is possible to keep modern populations in fair comfort on a small part of the working capacity of the modern world. If, at the end of the war, the scientific organisation, which had been created in order to liberate men for fighting and munition work had been preserved, and the hours of work had been cut down to four, all would have been well. Instead of that the old chaos was restored, those whose work was demanded were made to work long hours, and the rest were left to starve as unemployed . . .[24]

In *Player Piano* (Granada, 1977), Kurt Vonnegut Jr makes the same point about World War II, during which 'managers and engineers learnt to get along without their men and women, who went to fight. It was the miracle that won the war – production with almost no manpower . . . [It] was the know-how that won the war. Democracy owed its life to know-how.'

The work ethic is the Catch 22 of capitalism: our self-esteem and the opinion that others hold of us is measured almost entirely by what we *do* rather than what we *are*. Thus, an employed person of sixty-three is perceived (by himself as well as others) as being of much greater worth than he will be as an unemployed person of sixty-five; and a person who is forced into retirement by illness is immediately regarded as having lower status and value. This state of mind is also a major disincentive to early retirement, with its implication of 'age-ism' and 'statutory senility'.

The exaggeratedly absolute position asserts that work, however debased, is always good, while non-work, however welcome, is almost always bad, a form of incapacity and humiliation. It is as if work were seen as a raft in the middle of a shark-infested sea: being on the raft means safety and security, being in the sea means disaster; the idea of moving on and off the raft voluntarily has no appeal. There are winners and losers: no intermediate position is possible. The unemployed, invalids, old people, and women at home often feel rejected, and lack self-esteem and a sense of meaning and purpose in life because they do not work and are often 'put down' by those who do. Most people accept work as an essential part of their lives, giving them a sense of worth and achievement, and the workplace is an important element in their social relationships.

Work appears to be economically, socially, psychologically and perhaps even physiologically necessary for most people, and withdrawal from work – while welcomed by many – is dreaded by most. Yet while recognizing this, we should abandon the masochistic doctrine of work for work's sake. There is nothing inherently life-enhancing in performing boring and exhausting work year after year unless you actively prefer to do so. If the use of the bulldozer and traffic lights were banned in times of unemployment, many jobs in pick-and-shovel work and traffic direction could be created – but this would be socially pointless. We ought to welcome loosening, if not breaking, the chains that bind people to work.

For many people, unemployment has disastrous personal implications. Unemployment and leisure are opposite sides of the same coin – but psychologically their impact is totally different: one is feared, the other is eagerly sought. Unemployment is identified with rejection, uselessness, dependence on others, poverty, deprivation, the sense that the value of one's own personal time moves towards zero, and the elimination of the power to make significant personal choices. Leisure is sought and enjoyed, because it implies economic self-sufficiency and the power to make choices. Michael Young and Peter Willmott see 'work' and 'leisure' as being inextricably intertwined, each losing significance in the absence of the other. It is work, they say, that makes leisure meaningful, and vice versa. The need to work is analogous to the need to have a home, an essential element in self-definition: like the unemployed, homeless people may have a disabling loss of identity.[25]

We may be conditioned to accept the idea that work is psychologically necessary for all except a small, deviant minority – but the 'impressionistic' studies of people in work carried out by Studs Terkel (*Working*, 1975), Philip Toynbee (*A Working Life*, 1971), and Robert Fraser *et al.* (*Work*, 1968) suggest that many are kept on the treadmill merely by fear of losing their livelihood. Industrial democracy is based on faith in human capacity and diversity: workers should have more power to choose how and/or if they work.

SOCIAL ATTITUDES TO WORK AND RETIREMENT

Attitudes to work and retirement are largely matters of social conditioning. In Japan, a retirement age of fifty-five for men was set in the 1920s. This did not reflect the generosity of Japanese employers: it was at a time when the average male expectation of life was less than forty years. Since World War II, life expectancy in Japan has risen to seventy-six years for men (second only to Iceland) but the established retirement age – extended to fifty-seven years in some cases – remains for people in the lower ranks of Japan's industrial army (although some older people find work as cleaners or watchmen). A Japanese male of sixty, however, would not appear as an unemployment statistic.[26] Female participation in the Japanese labour force, although increasing, is lower than in Australia. In addition, many Japanese are employed as 'window-gazers', remaining on company payrolls as part of the paternalist tradition of 'lifetime employment' (see Chapter 6). Japanese management views this pragmatically: if the official unemployment figures are only 2 per cent, confidence in Japan's economy remains high, investment and consumption levels increase, and there is economic stability. If the 'window-gazers' were dismissed and put on relief, unemployment figures would treble overnight and citizens would perceive the economy as headed for serious decline.

Furthermore, like Italians and Greeks, the Japanese have a traditional three-generation family structure. It is taken for granted that when grandfather retires he will take an active part in the nurture of his grandchildren, his own children being freed for work and tasks outside the family. This model is not necessarily appropriate for export, however, and there is some evidence that the standard nuclear family of Australia, the United States and the UK is now the preferred model for young Japanese.

In Great Britain, *de facto* work sharing has evolved (apparently spontaneously) in specific employment areas such as the Post Office, British Rail, British Steel and British Leyland: productivity is low and wages even lower. However, there appears to be a consensus that it is better to adopt a co-operative approach to work and wages – in which, for example, four people share £200 per week (and with a lower output) – than to employ three for a higher wage and have one unemployed (with a higher output). This system has not been deliberately worked out on a national scale by trade-union leaders, but rather has evolved gradually in individual workplaces – although it was reinforced by the Employment Protection (Consolidation) Act of 1978 which made it very difficult to dismiss employees after six months' employment.

If the test of strict cost-efficiency was applied strictly to service employment in Britain, it has been estimated that it might decline by as much as 40–50 per cent. Britain appears to have adopted the philosophy that if you give a thousand people 1000 brooms and promise to pay them as long as they keep sweeping, this is preferable in social terms to having 100 workers operating complex cleaning equipment and putting 900 on the dole: employment is recognized as a form of social welfare. There has been a significant change in attitudes to work in Britain since the period between 13 December 1973 and 11 March 1974 when Edward Heath's Conservative government imposed a three-day working week, as an emergency response to fuel shortages and the national coal strike. It was expected that the three-day week would cut output by 40 per cent and apply irresistible pressure to the strikers to give in, but to everyone's surprise there was little (if any) fall in production and in some factories output actually increased. This raised the question of why, if the demand for goods and services can be met by three days' work instead of five, the two additional days should be worked.

If, as predicted by optimistic futurologists, within twenty years only 10 per cent of the labour force will be needed to produce all goods

IQ and an age of leisure?

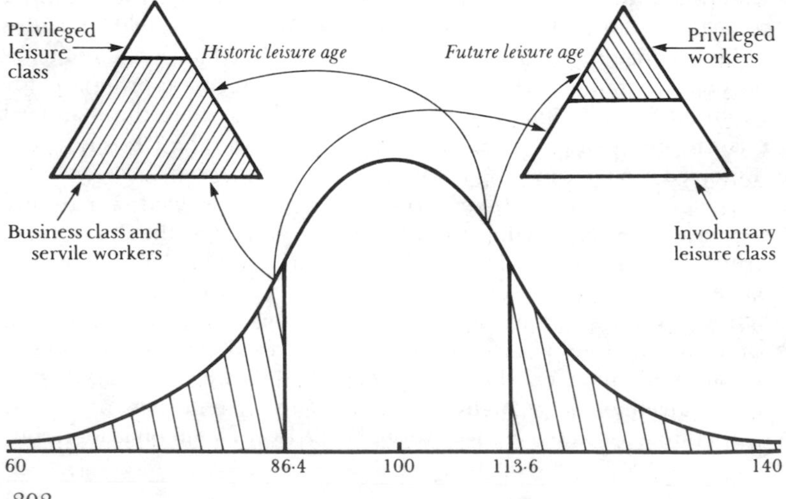

and services (making full allowance for rising levels of demand), will this be a Good Thing? a Bad Thing? or Morally Neutral? What will replace the traditional work ethic if only a minority need to work? In past golden ages of leisure – including Periclean Athens (as already described), some Italian city states during the Renaissance, France under Louis XIV, England and the Netherlands for much of the eighteenth century – the privileged classes pursued the arts, travelled, hunted, built, gardened, and discussed politics and history. The 'lower orders' supported them by providing routine services. We may face the problem of adjusting to an inversion of that historic model – a social pyramid of which the top part largely comprises a skilled meritocracy which works hard for high wages, and the lower part consists of the unskilled and semi-skilled who live in a condition of involuntary and unsought leisure because routine and repetitive work has been eliminated or significantly reduced. As James Martin and Adrian Norman put it, it seems likely that

> in general there will be a vast increase in demand for personnel of high IQ, logical ability and technical aptitude; as well there will be a shortage of the routine jobs in which the majority of mankind has been employed so far . . .
>
> A large proportion of the community today spends its long hours in routine. Take away from the clerk or the factory hand his routine work and he will be all at sea. To survive in a world without routine he will have to be creative, innovative, imaginative and adaptable; he must welcome innovation rather than shun it.[27]

Where are the schools that teach these skills? They do exist, but few of them are in the public sector and even fewer are found in the areas where low-income earners live and work. The elimination of routine work which could be carried out by inexperienced people is already well advanced, and since 1977 the unemployment rate of young Australians in the 15–19 age group has varied between 16 and 22 per cent of the civilian labour force in that age group.

Workers face an unpleasant dilemma. Technological change is adding significantly to unemployment for the poorly educated, but existing modes of employment continue to exploit a fair-sized unskilled proletariat in unpleasant jobs. About one worker in five is still physically disadvantaged by working conditions – on assembly lines, afflicted by excessive noise, heat or fumes, inhaling asbestos, digging underground, lifting heavy weights. Should such jobs be done by machines? Yes, but we must reject the facile optimism that suggests that all displaced workers will find new and agreeable jobs in the brave new world. We appear to be exploiting much of the working class and unemploying an increasing part of it simultaneously. Post-industrial technology can provide enormous increases in output and raise consumption levels appreciably while decreasing the need for a large labour force. It is essentially a matter of choice whether the results of this change provide hardship or benefit for society. If we adopt a high-productivity plus low-employment mix, it will be necessary to provide economic, social and psychologically satisfying alternatives to work for the less gifted. This will involve the following moves on the part of society:

1 Recognition that work need no longer be the primary mechanism for the redistribution of wealth.

2 Education based on the personal needs of each individual rather than the industrial needs of the community.

3 Encouraging individuals to recognize the value of individually determined time use.

4 Developing new forms of participation and recognition.

None of these alternatives will be easy. But they are inescapable, and must be tackled immediately.

GUARANTEED INCOME AND NATIONAL SUPERANNUATION

There are well-recognized exceptions to the rule that income is distributed by way of payment for work done – e.g. various social benefits and support schemes. This principle should be extended by adopting a 'guaranteed income' policy, as recommended by the Australian Poverty Commission; a national superannuation scheme would also provide people with a real choice between continuing at work or going into a voluntary retirement.

The 'poverty line' determined by the Poverty Commission in 1973 (see Chapter 6) varied according to the number of persons in the appropriate 'income unit' (generally families).[28] For a single person, the poverty line was 30·1 per cent of average weekly earnings if he was working and 24·3 per cent if he was not. For a family of two adults and seven children, the line was austerely set at 95·9 per cent of average weekly earnings if the head of the unit was working, and 90·3 per cent when he was not.

The commission recommended the introduction of a 'guaranteed income scheme' (also described as 'guaranteed minimum income', or GMI) to provide support for those below the poverty line – and up to 20 per cent above. This was considered preferable to piecemeal adjustments of existing support schemes – child endowment, family and disability allowances, unemployment benefits, students' allowances, aged and widow's pensions (service pensions were excluded from consideration). Guaranteed income would have been paid to all citizens – paupers and millionaires alike – as with the existing child-endowment scheme, financed by a proportional tax on *all* private incomes; the unemployed would not have been subject to a 'work test'. In effect, it would have been a form of 'negative income tax' – a scheme with impeccable conservative credentials, having first been proposed in the United States by Milton Friedman.[29] The Henderson Report was published in April 1975, the Whitlam government was removed in November 1975, and nothing whatever has been done to implement 'guaranteed minimum income'. Meanwhile, unemployment has tripled on CES figures (and increased sixfold if concealed unemployment and under-employment are considered). The Australian Labor Party has adopted the principle of guaranteed minimum income, but without specifying the exact form – or cost – of future implementation. The Whitlam government also set up a committee of inquiry into a national superannuation scheme, chaired by Professor K. J. Hancock of Flinders University. In 1976 the committee recommended, by majority, that a universal contributory scheme be established. This proposal aroused fierce resistance from the insurance industry, which saw such a scheme as a threat to their jobs, and the Fraser government rejected the Hancock Report (1979). The Australian Council for Trade Unions (ACTU) has adopted a national superannuation scheme as policy, and the powerful Storemen and Packers' Union has promoted the idea of a scheme along Swedish lines. These schemes have aroused strong opposition within industry, because a large superannuation fund which invested in business corporations could exert a strong influence on corporate policy-making.

RECOGNIZING TIME-USE VALUE OUTSIDE WORK

Since about 1970, the compulsory work ethic has been of declining significance to the educated, the thoughtful and the young. Its elimination will be a positive contribution to mental health. As discussed already, society should by all means recognize, encourage, reward and applaud work, but it should not punish its non-practitioners, especially as society grows richer while the labour force contracts. There needs to be a more flexible attitude towards non-work, to replace the 'social death' syndrome that now exists. People should be encouraged to recognize the personal value of time in their own lives, and to work out individual priorities. People – in or out of the labour force – should assert their freedom of choice, setting their own patterns of time use and learning to see leisure as an opportunity to fulfil their own potential. Freedom, however, means the right to make choices – and history abounds with examples of people preferring to accept life

Source: Drawing by Levin © 1979. The New Yorker Magazine, Inc.

patterns set or imposed by others rather than exercising their own right to choose. (The deaths of 914 people in Jonestown, Guyana in November 1978, followers of the Reverend Jim Jones is an appalling recent example.)

William Lambert Gardiner argues that modern society conditions people to overvalue 'extrinsic worth' – measured by employment and the acquisition of material possessions – and to undervalue 'intrinsic worth'.

> We are confronted with our own emptiness. Extrinsic motivation has destroyed intrinsic motivation . . . A disproportionate number of people die shortly after retirement. They are so conditioned to see themselves as an interchangeable part of a system that, when declared obsolete, they self-destruct. You can't use your spare time to gain intrinsic worth but you can use your spare money to gain extrinsic worth. You are compensated for your lifetime with money and you use that money in a vain effort to buy it back.[30]

For workaholics and others psychologically dependent on employment, therapeutic work may have to be found in 'welfare industries' dedicated to work for work's sake (without any economic justification).[31]

The work ethic is becoming increasingly anachronistic, as employees face an unequal competition with computers and other automated devices – which do not need to respond to words like 'ought' or 'duty' or the command to 'Work harder!'.

CHANGES IN THE WORKING YEAR

The typical Australian working year of 40 hours for forty-eight weeks totals 1920 hours, or 21·9 per cent of each year (8760 hours). If Australians worked 35 hours for forty-four weeks each year, there would be a fall of 20 per cent in total hours worked (1540). In theory, this ought to mean a potential increase of 20 per cent in the labour force, but in practice it seems unlikely. In many cases reduced hours in the working week have not altered productivity, and in some cases they have increased it. (Britain's productivity was 5 per cent less in 1980 than during Heath's three-day week in 1973–74.)

Australia has a shorter working year – about 229 days (one day short of forty-six working weeks) – than any other economically advanced nation. Four weeks' annual leave is normal and – in addition to one-day paid public holidays such as Labour Day, Anzac Day and Australia Day (often part of a long weekend) – there are several days of paid public holidays at Christmas, Easter and New Year. The average of hours actually worked each week in Australia exceeds forty. In 1972 the average in non-agricultural activities was 42·9 hours for males, 39·3 for females. Where industrial awards have provided for a 'standard' 35-hour week (e.g. in the petrochemical industry) many employees actually remain on duty for as many as forty-four hours because they prefer extra income to extra leisure (thirty-five hours are paid at normal award rates, the additional hours at penalty rates). Many workers also engage in second jobs, often within the 'informal economy', to escape taxation. About 40 per cent of Australian workers (teachers, public servants, office employees) already have a working week of 35 to 36.5 hours.

The 40-hour week remains the norm in most advanced economies. It was adopted in the US in 1933 by the National Industrial Recovery Act, and confirmed in 1938 by the Fair Labor Standards Act; by 1968, 37·1 hours were worked on average in manufacturing industry. In 1936 the 40-hour week was adopted in France (but up to twenty hours overtime was permitted) and in New Zealand, followed by Australia (1947), Canada (1972) and the Federal Republic of Germany (1974).[32] The USSR adopted a five-day week in 1967, and by 1971 an average of 40·4 hours were worked. In the United Kingdom the 40-hour week was 'standard' by 1966, but the average time worked was forty-five hours for men and 37·9 hours for women in 1972. In Japan in 1973,

42·3 hours were worked each week on average. In Switzerland an annual average (i.e. taking account of holidays) of 44·3 hours per week was worked. In Sweden the monthly hours worked in 1971 were 148 (i.e. thirty-seven hours a week), while Norway had Europe's lowest working hours, 32·7 for males and 28·5 for females (1973).[33] In 1981 the Netherlands and Belgium will adopt the 36-hour week.

Until 1980 there was relatively little union pressure to reduce the Australian working week below forty hours. This was a partial consequence of the phenomenon of urban sprawl, in that it took an inordinately long time to travel to and from work. (In future, time spent in commuting could well exceed the period spent at work.) Many employees prefer, once they have arrived at office or factory, to stay there for eight hours or so and accumulate the benefits of overall work reduction in the form of usable slabs of time – a shorter working year, or a shorter working lifetime.

On balance, a small reduction in hours worked each week (even the three-day weekend) is unlikely to create many new jobs. If that is so, the Australian Bank Employees Union's campaign for a 30-hour week seems unlikely to reverse the decline in bank employment – the small additional weekly time-gap can be filled more cheaply by mini-computers than by more workers.[34] The American Federation of Labor considered that for effective job creation the working week would need to fall to twenty-five hours.

The EEC has set a target of four weeks' annual paid holiday. The United Kingdom has no generally applicable statutory provisions for annual holidays with pay, entitlements being fixed by collective agreements or council orders (minimum entitlements are generally between two and four weeks). President Mitterrand introduced five weeks' annual leave in France in 1981, whilst West Germany provides only eighteen days. For those on a five-day week, the Netherlands, Ireland, Portugal and Spain provide fifteen days' paid leave; Luxemburg provides twenty-five days, and Italy ten (rising to fifteen days after 5 years' service, and to thirty days after 25 years). In many countries, leave is dependent on the number of years in service. In the United States, leave depends on labour agreements within particular industries. Two weeks' annual leave is normal for up to five years of service, four weeks after 18 years, and five weeks after 20 years. Canada grants two weeks' paid holiday after one year's employment, three weeks after 6 years, and four weeks after 10 years. Japan has seven days of paid leave after one year's service, eleven days after 5 years, sixteen days after 10 years, and twenty days after 15 years.

REDISTRIBUTING WORK: THE 35-YEAR WORKING LIFETIME

The greatest contribution to work redistribution in a post-industrial or post-service era would be to accept a 35-year working lifetime as the norm rather than the 50-year model which is still regarded as typical for working-class males – although not for women or the tertiary-educated. Eligibility for maximum retirement benefit could begin after thirty-five years' work (or less if medical or other special circumstances are involved). Since those who choose to remain working after this period will benefit by continuing to receive wages for a longer period, this might lead to a proportional *reduction* of retirement benefits where a choice is made to work forty, fifty or even sixty years in exceptional cases.

The basic 50-year to 35-year reduction would not occur immediately, but could and should be phased in to coincide with the reduced demand for labour which may accompany the increased use of labour-displacing technology. This would enable technology to realize the hope promised, but never fulfilled, by the Industrial Revolution: to liberate humanity from boring and life-denying labour. For those who found it psychologically necessary to continue working even past thirty-five years, even at tasks which most of us regarded as boring drudgery, the option must be kept open. Early retirement should not be made compulsory – because this would simply exchange the problem of youth unemployment for the problem of aged unemployment. Retirement at the age of fifty-five works well enough in Japan because the Japanese family structure adjusts to it. In Australia, it may cause enormous psychological problems of boredom and frustration – unless the concept of recurrent education can be made to work, and unless people begin to value free time for its own sake.

The traditional work ethic asserts that work means income and the power to make choices, and that free time means impotence and rejection. Often the reverse is true, and free time represents power to make choices whereas work means response to economic necessity – performing tasks set by someone else. The 35-year option may help people to conclude that it is better to make active choices on their own behalf rather than passively accept decisions made by someone else *for* them. The question of life after death has always occupied human thought: basic changes in human working patterns may stimulate interest in the possibility of life *before* death.

10 Technological Determinism: 'But We Have No Choice'

And there's a dreadful law here—it was made by mistake, but there it is—that if any one asks for machinery they have to have it and keep on using it.

The Magic City by Edith Nesbit

The term 'technological determinism' was coined by Thorstein Veblen to describe a society where basic decisions were shaped by available technological capacity, rather than the traditional political process based on ideology and value systems. In 1919 a Californian engineer, William Henry Smyth, invented the word 'technocracy' to describe 'the rule of the people made effective through the agency of their servants, the scientists and technicians'. In the 1920s Veblen and Howard Scott used 'technocracy' in another sense – government according to strictly rational scientific and/or technological principles, as a preferred alternative to what they regarded as the short-sighted and self-interested democratic system. The word 'technocrat' is often used pejoratively to describe those who promote decisions essentially based on technological capacity rather than human needs. But technologically determined decisions are not things that just happen. They are contrived, pushed and promoted by conscious human agencies – specialists in particular fields, many bureaucrats, advertising agents, manufacturers, newspapers and television – people who argue a position and, in default of any effective alternative view being put, win the debate.

Policy decisions are increasingly dominated by 'technological determinism': we face the possibility of rule by technocrats. There is an increasingly fatalist conviction that the answer to every complex problem is to be found in a technological 'fix' – and the more complex the 'fix', the more likely it is to be accepted without debate. The range and scope of political argument in technically complex areas is diminished – indeed, it verges on being irrelevant. We are often told: 'We have no choice but to mine uranium . . .', 'We have no choice but to build urban freeways . . .', 'We have no choice but to accept that per capita fuel consumption will double every ten years . . .'. If there *is* no option – and no turning back – then we might as well abandon any attempt to debate the issue or look for alternatives. A false dichotomy is forced on us which says that technology equals rationality and, therefore, opposition to (or proposals to modify) technology equals irrationality. It must be conceded at once that there are many ignorant and hysterical arguments in opposition to science and/or technology.[1] The basic questions which must be examined are: 'Must society be shaped by the available technology, or may society shape technology?' and 'Is technology a monolith, or are there varieties of technologies and are we free to choose between them?'. Is it inevitable that we adopt the highest technology available, irrespective of any adverse social consequences, such as mass unemployment, rapid resource depletion or the risk of atomic war?

The Judaeo-Christian tradition which has had such decisive influence on Western civilization advanced two different teachings about man's relationship with nature, each receiving about equal space in the Bible: (a) Man sharing with God transcendence over nature and transforming it; but also (b) Man as the good steward and trustee of nature, with a duty to tend the garden for all succeeding generations. The first view contributed to the nineteenth-century doctrine of material progress in which belief in God was replaced by a belief that science and truth were synonymous, and that technology was the pathway to solving human problems. In addition there was the

> curious assumption that nature is not only subservient and indestructible but also . . . incompetent. Industrial man sees himself as . . . forced to do, transform and convert because nature makes such a mess of it . . . Further, in modern economic accounting we speak of transformation as 'value added' . . . [and] all transformation is deemed useful.[2]

The second view is relied on by conservationists who urge their followers to live with nature rather than transforming it. Thomas Carlyle described man as a 'tool-making animal' and Karl Marx placed the material instruments of production at the central place in human development. These views underpin the idea of technological determinism, and the concept that we are most ourselves when we use tools. But is it the case?

I agree with Lewis Mumford's view that society has grossly misread the history of man's social development. We accept too readily the idea that man is primarily and essentially a tool-making animal: the oldest evidence of human society is found in stone hammers, axes and split skulls – there is no evidence of language, myth, religion, music, ritual, or magic to be dug up by the archaeologist's spade. We assume, all too readily, that the utilitarian and tangible are central to human experience and that emotion, communication, learning and understanding are peripheral. There is a fundamental dichotomy between inner man (the thinker and language-user) and outer man (the consumer and tool-user). When the amazing cave paintings of Altamira in northern Spain were found in 1879, it was assumed that they must be modern fakes: primitive man, the tool-user, could never (it was argued) have attempted anything so subtle, imaginative and moving. But, as Mumford says, if the only surviving evidence of Shakespeare was his lower jaw and fragments of his wife's crockery, we could hardly reconstruct *Hamlet*, *Twelfth Night*, or the sonnets from such materials.

> No single trait, not even tool-making, is sufficient to identify man . . .
>
> Man is pre-eminently a mind-making, self-mastering, and self-designing animal; the primary locus of all his activities lies first in his own organism, and in the social organization through which it finds fuller expression . . .[3]

Using contemporary man, not primitive man, as his starting point, Jacques Ellul wrote *La Technique ou L'enjeu du siècle* ('Technique, or the Stake of the Century') in 1954. This was revised and published in English in 1964 as *The Technological Society* (translated by John Wilkinson). Ellul defines *la technique* (not technology) as 'the totality of methods rationally arrived at and having absolute efficiency (for a given state of development) in every field of human activity'. As a result, 'Technical Man is fascinated by results, by the immediate consequences of setting standardized devices into motion, . . . committed to the never-ending search for "the one best way" to achieve any designated objective.' Robert K. Merton paraphrases Ellul's thesis:

> The technological society requires men to be content with what they are required to like; for those who are not content, it provides distractions – escape into absorption with technically dominated media of popular culture and communication and the process is a natural one: every part of a technical civilization responds to the social needs generated by technique itself. Progress then consists in progressive de-humanization – a busy, pointless, and in the end, suicidal submission to technique. The essential point . . . is that technique produces all this without a plan; no one wills it or arranges that it be so. Our technical civilization does not result from a Machiavellian scheme. It is a response to the 'laws of development' or technique.[4]

Necessity or freedom?

Ellul argues that technique is entirely anthropomorphic because human beings have become thoroughly technomorphic. Thus humanity is disoriented by technology, and turns away from its essence. The contemporary technical phenomenon has two essential characteristics – rationality and artificiality ('When we succeed in producing artificial *aurorae boreales*, night will disappear and perpetual day will reign over the planet').[5] These are supported by the automatism of technical choice ('The one best way' – something that can be measured empirically and is not subject to argument, eliminating the uncertainty of politics), self-augmentation (the concept of irreversible technical growth which develops according to geometric progression), monism (single-mindedness), universalism (the same phenomena appear all over the world, eliminating local particularism and tradition) and autonomy (technology can run itself and is less dependent on or related to people). Ellul is unremittingly pessimistic (but no more so than George Orwell):

> Freedom is not static but dynamic; not a vested interest, but a prize continually to be won. The moment man stops and resigns himself, he becomes subject to determinism. He is most enslaved when he thinks he is comfortably settled in freedom . . . In the modern world the most dangerous form of determinism is the technological phenomenon. It is not a question of getting rid of it, but, by an act of freedom, of transcending it . . . The first act of freedom is to become aware of necessity. The very fact that man can see, measure, and analyse the determinisms that press on him means that he can face them and, by so doing, act as a free man. If a man were to say: 'These are not necessities; I am free because of technique, or despite technique', this would prove he is totally determined. However, by grasping the real nature of the technological phenomenon, and the extent to which it is robbing him of freedom, he confronts the blind mechanisms as a conscious being . . .[6]

Langdon Winner has developed this theme with his term 'reverse adaptation' – the adjustment of human ends to match the character of available means.

> Abstract general ends – health, safety, comfort, nutrition, shelter, mobility, happiness, and so forth – become highly instrument-specific. The desire to move about becomes the desire to possess an automobile; the need to communicate becomes the necessity of having a telephone service; the need to eat becomes a need for a refrigerator, stove and convenient supermaket . . . Political reality becomes a set of institutions and practices shaped by the domination of technical requirements . . . Technology in a true sense *is legislation . . . is itself a phenomenon* . . . It is somnambulism (rather than determinism) that characterises technological politics – on the left, right and centre equally. Silence is its decisive mode of speech. If the founding fathers had slept through the convention in Philadelphia in 1787 and never uttered a word, their response to constitutional questions before them would have been similar to our own.[7]

The basic approach of determinists is to look at some major event, e.g. the French and Russian Revolutions, World Wars I and II, the impact of the motor car on urban environments, or the development of atomic weaponry, and say that it was inevitable – there was no other way it could have happened. But is it so? Car-driving and flying were born in the same decade. By a fateful social choice flying was made the subject of extreme, safety-conscious licensing and regulation while car-driving was let rip to become the big domestic killer of the twentieth century. Neither *had* to happen the way it did – we might have had safe roads and perilous skies. The tough regulation of shipping, safety and navigation (pioneered by James Plimsoll) began hesitantly in the 1840s and was much strengthened in the 1870s.

Increasing use of the car has been the dominant factor in shaping urban development in this century, as high-density living based around public-transport links was replaced by urban sprawl in many major cities. Low-density living was based on universal car use – and the lack of public transport led to increased dependence on private vehicles: the loss of a vehicle became a form of social crippling. Political pressure to build more freeways and arterial roads became irresistible. Los Angeles was the model 'car-based city of the future'. Its tramways and suburban railway systems were purchased by the car and rubber-tyre manufacturers, then eliminated and replaced by buses. Later the bus lines were abolished, and people came to accept – willingly enough – that the use of their own vehicles provided more flexibility and freedom of movement. More freeways generated more car dependence and greater population dispersal in increasingly far-flung suburbs: Los Angles, because of its low population density, could no longer be serviced by a comprehensive and profitable public-transport system. Its freeways were soon at capacity – and helicopters were considered as a form of mass transit. Its atmosphere was choked with fuel wastes, and any sense of community between urban neighbourhoods was destroyed as eight-lane freeways prevented pedestrian movement. This phenomenon was not inevitable – deliberate choices *were* made. By 1949, General Motors had taken a leading role in the replacement of more than 100 electric transit systems by GM buses in forty-five US cities, including New York, Philadelphia, Baltimore, St Louis, Oakland and Salt Lake City.[8] London and Paris are outstanding examples of choices being made the other way: public-transport grids were extended and co-ordinated so that the majority of vehicular journeys are now made by rail or bus. Neither of these cities attracts fewer residents or tourists than Los Angeles.

Many still accept, almost without question, that in Western society per capita energy consumption will double every decade, that this is a good thing in itself and that nothing can be done about it. Basically, every community determines the rate of energy usage that it *wants*: the USA and Canada use large quantities of energy – with enormous waste – partly because their cities use more private than public transport. European states maintaining a similar standard of living use far less energy – even in cold areas – because their houses have insulation, double glazing, the use of thermodynamically efficient 'heat pumps' and other fuel-saving measures. The state of Victoria, with a Mediterranean climate, has the same per capita energy usage as Finland with an eight-month winter. Tasmania has by far the highest per capita production of electricity in the world, despite – or perhaps because of? – its poor resource base (apart from water power). Tasmania generates 17 992 kWh per person per annum, compared to 9363 in the USA and 5240 in West Germany. Comalco uses 27 per cent of Tasmania's power and contributes 0.5 per cent of employment. The Hydro-Electricity Commission (HEC) is a powerful lobby group for increased electricity generation: Tasmania's energy policy appears to be based on the dubious assumption that generating cheap electricity will create more industrial jobs.[9] It is, however, safe to generalize that in all advanced economies as electricity generation increases, industrial employment falls: Tasmania's output has increased enormously since 1950, and the industrial labour force has fallen proportionally – but instead of re-examining the basic assumption, the HEC argues that the generation of even more power may do the trick.

European economy, Tasmanian extravagance, North American and other uranium developments are all the results of *choices* being made. But the relevant interests and lobby groups work hard to sell the idea that there is no choice, it is inevitable. The advocates of nuclear power argue that it is inevitable that the world will go nuclear. The abundance of uranium and the technology of the fast-breeder reactor are major disincentives to the development of non-polluting, non-exhaustible energy sources (e.g. solar, wind, tidal, geothermal). If there was no uranium at all, is it suggested that Western civilization would stagger to a halt? As the 1967 Nobel Prizewinner for Chemistry, Sir George Porter, wrote, 'If sunbeams were weapons of war, we would have had solar energy centuries ago.' Ralph Nader speculated that if sunlight was a scarce commodity, controlled by an enterprising syndicate which had built a huge screen across the earth's land surface, then solar energy would be the greatest of all industries: 'If Exxon owned the sun we would have had solar energy years ago. . .'. Technological determinism crosses all political boundaries. It has adherents among ardent Marxists and far-right capitalists – its support is least towards the centre and with the uncommitted.

The false premise on which technological determinism is based asserts that technology is a single entity, monolithic and incapable of being differentiated. There is no suggestion that there are varieties of technologies, or that is possible for nations to choose between them. This is the 'cargo cult' view of technology: we wake up one morning to find a computer in the garden, it has arrived impersonally and we must take it or leave it as we find it. Technological determinists argue that if we reject high technology we will be punished; if we accept it, the pre-recorded birds will sing all day, and artificial lighting will abolish night. It is important to distinguish between varieties of technologies:

1	Labour-displacing	versus	labour-complementing.
2	Megatechnics (e.g. nuclear energy)	versus	polytechnics (e.g. solar energy).
3	Centralized	versus	decentralized.
4	High entropic	versus	low entropic.
5	Environmentally harmful	versus	environmentally benign.

In *The Costs of Economic Growth*, E. J. Mishan tells 'a sort of parable' of a gun-dependent society in which everyone carried firearms, where shootings were a routine occurrence, all houses and offices were made bullet-proof, the manufacturers of armour and shatterproof glass were major employers, economists expressed gratification at the growth in gun-based and related industries, the sales of liquor and drugs were booming, hospitals and police created thousands of jobs and the funeral business was bullish. Mr B then suggests an improved technological model for city design at enormous expense, based on 'pistol architecture' which will enable people to keep their guns while reducing morbidity to acceptable levels. Since the gun is essential to the economy, nobody is foolish or reactionary enough to suggest that eliminating gun dependence might be a solution – after all, you can't turn back the clock.

But suppose that the speed limitations of the 1890s and the famous man with the red flag had remained in force in built-up areas, and this had led to the development of transport alternatives such as maintaining high-density cities with comprehensive public-transport grids. Suppose that by the 1950s, Australian and American cities had adopted the London or Paris option rather than the Los Angeles one – that 100 000 lives were saved annually, atmospheric pollution was a minor factor, the sense of community was encouraged, pedestrians could walk without fear, and city planning was based on the needs of the individual rather than the demands of the car. It may well be too late to change now – but was it always 'inevitable' that society had to sacrifice so many lives, so many torn bodies, so much city devastation to the Minotaur? Who made it compulsory?

'SPEER'S SYNDROME'

Technological determinism in its extreme form could appropriately be called 'Speer's Syndrome', after Albert Speer, a careerist technocrat who was Hitler's Minister of Armaments and Munitions from 1942 to 1945. He wrote:

> We owed the success of our program to thousands of technicians with special achievements to their credit to whom we now entrusted the responsibility for whole segments of the armaments industry. This aroused their buried enthusiasm . . . Basically, I exploited the phenomenon of the technician's often blind devotion to his task. Because of what seems to be the moral neutrality of technology, these people were without scruples about their activities. The more technical the world imposed on us by the war, the more dangerous was this indifference of the technician to the direct consequence of his anonymous activities . . .
>
> The criminal events of these years were not only an outgrowth of Hitler's personality. The extent of the crimes was also due to the fact that Hitler was the first to be able to employ the implements of technology to multiply crime.[10]

At his testimony during the Nuremberg Trials, Speer said: 'The more technological the world becomes, the greater is the danger . . . As the former minister in charge of a highly developed armaments economy it is my last duty to state: A great new war will end with the destruction of human culture and civilisation. There is nothing to stop unleashed technology and science from completing its work of destroying man which it has so terribly begun in this war.'

The scientists who worked on the atomic bomb were driven on by a sense of technological compulsion: the conviction that if the Allies did not make the bomb first, Hitler would; followed (after Germany's defeat) by a certainty that its use against Japan would save millions of Allied lives, together with a sense of awed exhilaration at having the opportunity to work on a scientific achievement of unparalleled importance. The politicians (Truman, Stimson) were persuaded that there was no option, and the first atomic bombs were dropped on Hiroshima and Nagasaki.

During the Cold War, scientists responded to the great intellectual challenge of the hydrogen bomb. Langdon Winner has pointed to an answer by Dr J. Robert Oppenheimer when he was asked by the Personal Security Board in 1954 whether his doubts about the bomb increased when it became clear that it could be made:

I think it is the opposite of true. Let us not say about use. But my feeling about development became quite different when the practicabilities became clear. When I saw how to do it, it was clear to me that one had to at least make the thing. Then the only problem was what one would do about them when one had them. The program in 1949 was a tortured thing that you could well argue did not make a great deal of technical sense. It was therefore possible to argue also that you did not want to even if you could have it. The program in 1951 was technically so sweet that you could not argue about it.[11]

Oppenheimer's remark was not just a slip of the tongue: elsewhere he said of the hydrogen bomb, 'From a technical point of view it was a sweet and lovely and beautiful job.'[12]

The mathematician and computer pioneer John von Neumann said, 'Technological possibilities are irresistible to man. If man *can* go to the moon, he *will*. If he can control the climate, he *will*.' (Mumford pointed out that the logical extension of this argument was: *If man has the power to exterminate all life on earth, he will.*) The American geneticist and Nobel Prizewinner Herman Muller (1890–1967) stated his views frankly:

Man as a whole must rise to became worthy of his best achievement. Unless the average man can understand the world that the scientists have discovered, unless he can learn to comprehend the techniques he now uses, and their remote and larger effects, unless he can enter into the thrill of being a conscious participant in the great human enterprise and find genuine fulfilment in playing a constructive part in it, he will fall into the position of an ever less important cog in a vast machine. In this situation, his own powers of determining his fate and his very will to do so will dwindle, and the minority who rule over him will eventually find ways of doing without him.[13]

Muller argued for genetic controls over human breeding by using sperm banks to which geniuses had made donations: 'How many women, in an enlightened community devoid of superstitious taboos and of sex slavery, would be eager and proud to bear and rear a child of Lenin or of Darwin.' He wrote of his proposed sperm banks, 'Their mere existence will finally result in an irresistible incentive to use them.' In 1980, a team in California directed by William B. Shockley was planning the conception and birth of a 'superchild' in accordance with Muller's ideas. Shockley, co-inventor of the transistor and a Nobel Prizewinner in Physics, was an active promoter of a 'sperm bank'. The concept is based on a fallacy. The sperm of a genius is of far less value than the sperm of a genius' father – the fathers of Shakespeare, Bach, Mozart, Churchill and Roosevelt produced more successful offspring than their famous sons. (See Roald Dahl's novel *My Uncle Oswald*, Penguin, 1980, for a witty treatment of this theme.)

In *The Pentagon of Power*, Mumford wrote:

Western society has accepted as unquestionable a technological imperative that is quite as arbitrary as the most primitive taboo: not merely the duty to foster invention and constantly to create technological novelties, but equally the duty to surrender to these novelties unconditionally, just because they are offered, without respect to their human consequences.[14]

He examined the postulates of a society based on a new industrial complex built around the 'megamachine':

There is only one efficient speed, *faster*; only one attractive destination, *farther away*; only one desirable size, *bigger*, only one rational quantitative goal, *more*. On these assumptions the object of human life, and therefore of the whole production mechanism, is to remove limits, to hasten the pace of change, to smooth out seasonal rhythms and reduce regional contrasts . . . Cultural accumulation and stability thus become stigmatised as signs of human backwardness and insufficiency.[15]

Joseph Weizenbaum has agonized about the impact of computers on decision-making. He sees 'a culture already enthralled by what economists call the pig principle: if something is good, more is better', and argues that 'the myth of technological and political and social inevitability is a powerful tranquiliser of the conscience. Its service is to remove responsibility from the shoulders of everyone who truly believes in it.' Weizenbaum argues that reason must be related to 'human dignity, to authenticity, to self-esteem, and to individual autonomy.'

The technologist . . . will try to construe all arguments against his megalomanic visions as being arguments for the abandonment of reason, rationality, science and technology, and in favour of pure intuition, drug-induced mindlessness, and so on. In fact I *am* arguing for rationality. But I argue that rationality may not be separated from intuition and feeling. I argue for the rational use of science and technology, not for its mystification, let alone its abandonment. I urge the introduction of ethical thought into science planning. I combat the imperialism of instrumental reason, not reason . . .

It also used to be said that religion was the opiate of the people. I suppose that saying meant that the people were drugged with visions of the good life that would surely be theirs if they but patiently endured the earthly hell their masters made for them. On the other hand, it may be that religion was not addictive at all. Had it been, perhaps God would not have died and the new rationality would not have won out over grace. But instrumental reason, triumphant technique, and unbridled science *are* addictive. They create a concrete reality, a self-fulfilling nightmare. The optimistic technologists may yet be right: perhaps we have reached the point of no return. But why is the crew that has taken us this far cheering? Why do the passengers not look up from their games? Finally, now that we and no longer God are playing dice with the universe, how do we keep from coming up craps?[16]

GROWTHMANIA

The concept of 'instrumental reason' is closely linked with the idea of growth for growth's sake. Not all growth is beneficial – cancer, for example. As E. F. Schumacher used to say, 'In my grandchildren, growth is a wonderful thing – but if I started to grow again at my age, it would be tragic.'

E. J. Mishan has attacked what he called 'growthmania' – the concept that more is necessarily better, and that a society where the GDP increases by 6 per cent each year must be happier and more satisfying to live in than a society which has 'lost the race' with a GDP growth rate of only 2 per cent. Suppose that it was established that Country X, with a 6 per cent GDP growth rate, also had rising figures for urban crime, pollution, suicide, alcoholism, drug dependence, mental breakdown, and deaths from cancer; while Country Y had only a 2 per cent growth rate, with lower indicators of social pathology. If the phenomena were related, would we still push for growth *at any cost?*

Great Britain has had a very poor press in recent years, with much alarm and despondency being generated over low GDP rates, a growing 'technology gap', and a sense that the nation is falling behind in an international, if not intergalactic, economic competition. The British sociologist Michael Young takes a different view. He argues that although Britain scores poorly in GDP annual growth rates, it scores highly in the following areas: longevity (now higher than Australia or the US), infant mortality, literacy, telephone use, access to public transport, proportion of population living in detached houses with gardens, political stability (Ulster notwithstanding), high-quality media and culture, unspoiled landscape, and relatively low rates for crime and drug addiction.[17] Is it better to live to seventy-five years in a civilized society with a 2 per cent growth rate, or to die at sixty in a ruthlessly competitive society with a 6 per cent growth rate? The growthmaniacs never examine this question.

Mishan and Young argue from different bases. Mishan is a conservative economist with basically environmental misgivings about growth, and is less concerned with economic equity or redistribution of wealth. Young, although a life peer, proposes socialist and egalitarian policies: he asserts in effect that the right choice of product, fairly and thoughtfully distributed, is much more valuable than mere aggregate product, especially if ill-chosen and mal-distributed.

The transfer and control of technology

In the Second and Third Industrial Revolutions there were two basic models of acquiring technology from its country of origin, the national and the colonial. The United States, Japan and Sweden adopted the national model, and are now technological exporters in their own right. When the United States built its railways in the 1860s, loans were raised, locomotives bought and engineers borrowed from Britain – but ownership remained in American hands. After World War II, Japan acquired licences to use British and American technology, modified it to meet local needs and produced Datsuns, Toyotas and Mazdas, rather than inviting American corporations to set up Ford or GM plants. Sweden, with a population of only 8 million and a low resource base, has developed its own technology in competition with Britain, the USA, Germany or Japan. As a result, Swedish high-technology companies – some of them multi-nationals – have won worldwide markets: Volvo (cars) and Saab-Scania (cars, trucks and aircraft), SKF (bearings), L. M. Ericsson (telecommunications equipment), Bofors (armaments), Electrolux (vacuum cleaners, robots and lawnmowers), Hasselblad (cameras), and ASEA (robots and engineering).[18]

Australia, like India, Mexico and much of South America, adopted the colonial model of technology transfer. Technology was developed abroad, its owners exported it to Australia, and its local manufacture was largely under foreign control. Australia continues to import technology and is an insignificant exporter. The primary agents of technology transfer in Australia were originally foreign-owned companies and later multi-national corporations. The Myers Report ignored the implications of foreign ownership, technology transfer and national sovereignty except for this isolated and apparently naive comment:

> [Para. 4.72] Technology is in general an international commodity and the ability of the Government or people of a small country such as Australia to influence the development of new technologies and their use in the world is probably extremely limited.[19]

Sweden, located close to large, rich markets, has two apparent disadvantages – its small population, and the fact that no other nation speaks Swedish. In fact, however, the Swedes were thereby forced to develop their own human resources. Australia has an apparent advantage – its role as part of the English-speaking world. In reality, this facilitates the penetration of the country with products and technologies from corporate interests in the USA and Britain, which contributes to a sense of complacency and/or pessimism: 'Why bother to invent the wheel? If it is any good, the Americans will sell it to us.'

Multi-national corporations dominate the commanding heights of the Australian economy – coal, copper, aluminium, uranium, motor manufacturing, food processing, computers, drugs, chemicals, plastics, petrol and advertising. Australia's media have done nothing to stimulate debate on the link between multi-national corporations and technology transfer. The twelve largest advertising agencies in Australia are all or partially owned by US interests. Together with the press and commercial television, they have decisive roles in encouraging technological determinism and making consumption-based life-styles seem compulsory (and rejection, or even questioning, of them is made to appear eccentric, irrational or even subversive). The media, with a few honourable exceptions, has shown a monumental lack of interest in, or even curiosity about, the impact of high technology on employment or society. Much of the press has exhibited a 'gee whiz' enthusiasm for technological hardware and software – not unrelated to the advertising revenue it generates. The so-called 'serious' press has shown an astonishing degree of gullibility in accepting the views of the technological determinists – credulous public servants, incurious journalists and vigorous salesmen – virtually without question.[20]

In Australia the multi-nationals pushed against an unlocked door, and were probably more powerful (and less subject to effective scrutiny or criticism) than anywhere but Canada. The prevailing colonial mentality discouraged any sense of Australia striking out for itself economically as Japan, Sweden or Switzerland had done, depressed research and development, made higher education seem pointless, and led to a 'brain drain' of able graduates to Britain or the USA. Australia's Foreign Investment Review Board (FIRB) considered 4437 proposals for foreign investment between April 1976 and December 1979, and rejected only thirty. Its Deputy Chairman, Sir William Pettingall, referring to the possibility that the Australian labour movement would prevent uranium exports, said that it was 'immoral' to withhold anything of value from people who wanted it and that they would be morally justified in using force to obtain what they wanted – an argument which sounded like a statement by counsel for the defence in a rape trial. The Whitlam government required 100 per cent Australian ownership of new uranium developments, but the Fraser government reduced this first to 75 per cent, then to 50 per cent. In 1978 the guidelines were further eased, and a company with 25 per cent or more Australian ownership may be treated as if it had 50 per cent.

Many influential Australians seemed more pleased than alarmed that this continent was seen only as a source of raw materials and were grateful when they were repaid with foreign technology. Sir Henry Bolte, as Liberal Premier of Victoria, was not ironical when he expressed the hope that Australia would become 'the world's quarry'. In *Will She Be Right? The Future of Australia* by Herman Kahn and Thomas Pepper, the last page contains this defensive comment: 'Some Japanese and Korean friends went so far as to express bewilderment as to why Hudson Institute would want to study a country they took to be as uninteresting and, except for its resources, as unimportant as Australia.'[21]

In its 1979 Annual Report, the FIRB stated that the amount of funds provided by Australian investors to foreign-controlled enterprises was significant and had been steadily increasing throughout the 1970s. Of the net increase in funds employed in 1976–77 by these enterprises, more than half was derived from Australian resources. It is estimated that in the eight years between 1969 and 1977, total Australian loans to these enterprises increased fivefold.

The phenomenon of Australians lending money to multi-nationals to enable them to acquire control of the Australian economy, because they are not prepared to take investment risks themselves, illustrates a fundamental diffidence in the Australian character. There is a basic hesitancy about local capital formation, and a fatalist sense that foreign take-overs are part of the natural order. It also confirms the impotence of Australia's political system, and the failure of its governments to grasp the nature and extent of economic and technological change. Australia had significantly lost control of its economic destiny by 1980. Profitability was high for large corporations, but there were record numbers of bankruptcies in middle and small businesses. Inflation was high, but so was demand: so too was unemployment, contrary to classical economic theory (see also Chapter 6). The professional economists were intellectually bankrupt; the political process had become enervated and virtually irrelevant as a factor in determining the shape of things to come.

Australia has the economic history and corporate profile of an ageing country, not a new one. R. W. Connell points out that of the fifty largest companies operating in Australia, only four (Chrysler 1951, Comalco 1960, Alcoa 1961, CRA 1962) – all of them foreign-owned – have been set up since 1936.[22] In the United States, where new industries such as television or computer chips were developed, the highest growth rates have come from companies which did not exist forty years ago. In Australia, the tendency is towards consolidation in already existing companies, and new industries are taken up by foreign companies. Motor manufacturing was absorbed by General Motors and Ford, and by Chrysler and then Mitsubishi, television was swallowed by newspapers, publishing corporations, and a transport company. Xerography and computers were fully imported by multinationals. The great mining boom of the 1960s led to the creation of new enterprises which were then taken over by foreign corporations. Food processing was dominated by foreign-owned interests.

The Senate Standing Committee on Science and the Environment reported that 'Of the royalty payments and payments for technical know-how by the business enterprise sector in 1973–74, 94 per cent was paid overseas – more than half to related foreign enterprises.'[23] The report noted that in terms of patents taken out each year per million people, Australia ranked a poor eighteenth of twenty-four advanced nations, and seventeenth of twenty-one nations for patents per $US1000 of GNP. (Switzerland came first in both cases.) The failure of the Myers Report to devote even a paragraph to the questions of who owns and controls technology, and whether Australia has the right to choose between technologies, is inexplicable and indefensible. Nevertheless it mirrors Australia's apathy, fatalism and lack of intellectual curiosity without much distortion.

But is there a choice?

Lewis Mumford has distinguished between 'megatechnics', which he defines as the concentration of power in a centrally controlled authoritarian system, and 'polytechnics', where power is decentralized in a pluralist, democratic manner. This dichotomy has been taken up by a number of writers including Paul Goodman, Ivan Illich, David Dickson, Robin Clarke, Jim Cairns, Kit Pedler, E. F. Schumacher, and the Canadian GAMMA team. All have argued for breaking away from the psychological grip of technological determinism. In *Small is Beautiful: a Study of Economics as if People Mattered*, (Abacus, London, 1973) Schumacher argued for reversing economies of scale. Instead of more things being produced by fewer people using high technology, it might be better for society to adopt 'Buddhist economics' – low-technology and high-employment techniques which would inflict the least violence on society or its environment. He urged that small-scale labour-intensive industries serving regional needs ought to be encouraged, instead of being squeezed out by large capital-intensive firms with huge overheads aimed at larger markets (e.g. 100 small bakeries are preferable to five large ones because the smaller units will absorb more workers). Because of the extreme centralization of technologically advanced economies, this option may not be really open (although it may be so in India, which Schumacher wrote much about). Many young people have opted out of a high-technology competitive urban lifestyle in favour of a low-technology non-competitive rural life influenced by Schumacher and his emphasis on 'appropriate technology'. These groups are largely self-supporting and comprise a relatively small proportion of the labour force.

Robin Clarke, of Biotechnic Research and Development (BRAD) in the United Kingdom, prepared a list of criteria distinguishing between 'high technology' (hard) and 'utopian technology' (soft), as set out in the table on p. 228.

One of the most important experiments in devising alternative appropriate technology has occurred in Britain. After massive introduction of high technology in the 1970s, Lucas Aerospace reduced its labour force from 18 000 to 12 000. The Lucas Aerospace Combine Shop Stewards' Committee reacted by producing a 'corporate plan' in which more than a hundred and fifty ideas for new products were proposed. All could be built with existing machine tools and skills, and were non-alienating, non-fragmenting and energy-saving.[24] The Lucas management resisted the plan on the basis that it threatened the traditional prerogatives of management. The new product proposals included the following:

1 A machine called the Hobcart to enable children with Spina Bifida to propel themselves round.

2 Portable low-cost, life-support systems to keep acute cardiac patients alive on their way to the hospital.

3 Gaseous hydrogen fuel cells to conserve energy and avoid such anomalies as, for example, using more energy to cool New York in summer than to heat it in winter.

4 A hybrid power-pack for vehicles which could combine the take-off torque of petrol engines and the long-distance performance characteristics of electric motors.

5 Low-cost portable kidney dialysis machines which would save 3000 lives each year in Britain alone.

6 Telechiric ('hands at a distance') devices, permitting workers to use and develop personal skills in controlling sensitive but dangerous processes.

7 Cheap solar-powered irrigation pumps.

8 Bicycle-operated machinery for grinding (for use in Third World countries).

9 Heat pumps: low-cost and low-energy-consuming 'heat exchangers' which take heat from the atmosphere and transfer it indoors (reversing the refrigeration cycle).

10 A hybrid road and rail vehicle: a bus fitted with railway bogies so that it could run on tracks as well as tyres. Developed in Britain at a

total cost of £10 000, this could have an enormous impact on urban transport systems, in integrating fixed rail systems with catchment areas dependent on roads, and saving on fuel consumption and road usage.[25]

Mike Cooley, the guiding spirit of the plan, has pointed to the absurdity that Lucas workers were employed in making refinements to the supersonic Concorde at a time when thousands of old people in Britain were dying of hypothermia because they could not afford heating; and that, for all the increased sophistication of modern vehicles, the average speed of vehicles in London's streets had fallen from 18 kph in 1900 to 10 kph in 1980. The committee was nominated by 180 organizations for the 1979 Nobel Peace Prize. In the United States, Congressman Clarence D. Long (Democrat, Maryland) has actively promoted the idea of 'light-capital technology', intended to make 'the most effective use of available, renewable materials and legal "know-how", while meeting the end-use needs of the community'. He attempted to amend the Humphrey-Hawkins (Full Employment and Balanced Growth) Act of 1978 by adding a new section, 'Appropriate and Light Capital Technologies', which are defined as technologies which:

(1) are small in scale, simple to install, and durable in operation;
(2) are labor rather than capital intensive;
(3) are not dependent on a highly centralized infrastructure for production, maintenance, or repair;
(4) make effective, efficient use of available and particularly of renewable resources;
(5) meet the needs of local communities and enhance the self-reliance of such communities; and
(6) enhance rather than degrade the environment.

It was estimated that in the Nassau-Suffolk area of Long Island, New York, for example, a light-capital energy programme aimed at installing insulation, storm windows and solar hot-water systems could provide over 250 per cent more employment and 200 per cent more usable energy over a thirty-year period than the same amount of money spent on constructing and maintaining nuclear power plants. Long has argued that much high technology sent to Latin America and Africa as part of US foreign-aid programmes has been utterly inappropriate to local needs, reflecting the technological determinism of the donors rather than the needs of the recipients.

It is slowly being recognized that in many cases small may be appropriate. Many large corporations now concede that greater social and economic efficiency can result from breaking down manufacturing into smaller units: two-thirds of job creation in the US between 1970 and 1980 was in firms with less than twenty employees. The adoption of blockbuster technologies – e.g. converting a river valley into a dam to provide industrial power – may provide apparent economic advantages in the short term, but lead to long-term dis-economies. It may be more desirable to have five small community-based hospitals than one megahospital, or to devolve administration rather than centralize it. We need far greater flexibility in looking for the proper relationship between the use of capital, labour, raw materials and land. As Shirley Williams says, instead of asking 'What is the return per worker employed?, we should ask 'What is the return to all the factors of production taken together?' It is also essential that we break the old link between highly wasteful energy growth and the growth of output – using vast quantities of prime fuel to produce small quantities of energy at the point of use (and wasting the difference by releasing heat into the atmosphere).[26]

The GAMMA Report was critical of *The Limits to Growth* (1972), a neo-Malthusian report on prospects for future world development commissioned by the Club of Rome, and written by Donella and Dennis Meadows with a team of systems dynamicists from MIT. Many of the 'Doomsday' predictions of *The Limits to Growth* have been invalidated by discoveries of vast reserves of raw materials in the 1970s. GAMMA rejected the concept of 'limits to growth' and instead proposed 'limits to forward throughput', arguing that more attention had to be given to *means* rather than *ends*. If a customer buys a container of soft drink, there are significant differences in the ecological, economic and environmental consequences of choosing particular containers – e.g. aluminium cans (non-recyclable), aluminium cans (recyclable), steel cans (generally non-recyclable but using far less energy than aluminium cans), 'non-returnable' glass bottles or returnable bottles. The choice of gas or electricity for cooking has serious implications: natural gas is burnt to drive electric generators which deliver electricity to the kitchen, a process which involves degrading the potential of the original fuel and adds to pollution and entropy. Most urban transport is designed to maximize private vehicle movement, an extraordinarily wasteful use of resources which actually inhibits personal mobility (see also Chapter 4). The desired ends – soft-drink consumption, cooking, or personal mobility – may be worthwhile in themselves; but do we minimize or maximize the use of resources which lead to the ends?

The GAMMA Report put its greatest emphasis on conservation, the promotion of a sense of community, developing intrinsic rather than

Characteristics of 'hard' and 'soft' technology

Hard technology society	Soft technology society
Ecologically unsound	Ecologically sound
Large energy input	Small energy input
High pollution rate	Low or no pollution rate
Non-reversible use of materials and energy sources	Reversible materials and energy sources only
Functional for limited time only	Functional for all time
Mass production	Craft industry
High specialization	Low specialization
Nuclear family	Communal units
City emphasis	Village emphasis
Alienation from nature	Integration with nature
Consensus politics	Democratic politics
Technical boundaries set by wealth	Technical boundaries set by nature
Worldwide trade	Local bartering
Destructive of local culture	Compatible with local culture
Technology liable to misuse	Safeguards against misuse
Highly destructive to other species	Depends on well-being of other species
Innovation regulated by profit and war	Innovation regulated by need
Growth-oriented economy	Steady-state economy
Capital-intensive	Labour-intensive
Alienates young and old	Integrates young and old
Centralist	Decentralist
General efficiency increases with size	General efficiency increases with smallness
Operating modes too complicated for general comprehension	Operating modes understandable by all
Technological accidents frequent and serious	Technological accidents few and unimportant
Singular solutions to technical and social problems	Diverse solutions to technical and social problems
Agricultural emphasis on monoculture	Agricultural emphasis on diversity
Quantity criteria highly valued	Quality criteria highly valued
Food production a specialized industry	Food production shared by all
Work undertaken primarily for income	Work undertaken primarily for satisfaction
Small units totally dependent on others	Small units self-sufficient
Science and technology alienated from culture	Science and technology integrated with culture
Science and technology performed by specialist elites	Science and technology performed by all
Strong work/leisure distinction	Weak or non-existent work/leisure distinction
High unemployment	(Concept not valid)
Technical goals valid for only a proportion of the globe for a finite time	Technical goals valid for all people for all time

Source: Robin Clarke, 'Biotechnical Research and Development (Britain)' reprinted in David Dickson. *Alternative Technology* (Fontana Press, 1974).

extrinsic values (e.g. through education), reduced commodity dependence, and a critical approach to artificially stimulated needs. It does not examine the degree to which waste and inefficiency are promoters of employment, as I have argued: instead it concentrates on the need to eliminate waste as a major source of high entropy. There is comparatively little discussion of the impact of technology on employment, expecially in computer-based industries.

GAMMA defined the 'Selective Conserver Society' (see Chapter 6) as one which strives to 'a. conserve imputs and outputs by reducing waste wherever and whenever it is encountered b. develop an increasing harmony with, rather than opposition to, Nature and c. measure the long term consequences of its acts *before* undertaking them.' Like Australia, Canada has a resources-based economy and the 1970s were marked by spectacular discoveries of oils and metals. The report argued for limits to urban growth – an option which is more open in Canada than in Australia – and endorsed Schumacher's advocacy of 'alternative technology'. It made many policy recommendations, including the creation of a Canadian conservation corporation to initiate a national public debate on the 'conserver society' and to investigate ways of conserving resources. (It also stressed Canada's moral responsibility to promote this concept internationally.) The report has been surprisingly uncontroversial in its impact. Canada's three main political parties have adopted a characteristic compromise – they accepted the concept of the conserver society but have done nothing whatever to implement it. This has reduced debate to a minimum.

In 1977 a New Zealand Planning Council and a Commission for the Future were established by legislation.[27] The first is to 'act as a focal point for a process of consultative planning about New Zealand's medium-term development', and to assist in promoting discussion on economic, environmental, social and cultural issues. The second body is concerned with 'long-term economic and social development', with special attention being given to prospective developments in science and technology. The Commission for the Future has published a series of short, lively papers and taken an active role in stimulating a 'futures' debate. The legislation provides that members shall include a minister, an Opposition MP and the Director of the Department of Scientific and Industrial Research. There is no Australian equivalent: the Australian Science and Technology Council (ASTEC), set up in 1979, is concerned with *how* things happen not *why*, and relates science and technology to an industrial context. Its reports are bereft of social or philosophical concerns.

The New Zealand Commission for the Future authorized the

National Research Bureau to conduct a survey of 2000 people aged fifteen years and over. They were asked detailed questions about national and personal goals, and to choose between four different scenarios.

Option 1: High growth in economic living standards is encouraged, with no effort to improve social and environmental conditions, the aim being for future economic living standards to have increased sufficiently to offset possible worsening social and environmental living standards.

Option 2: Moderate growth in economic living standards is encouraged, with economic, social and human resources also being used to encourage limited improvements in social and environmental conditions.

Option 3: Growth in economic living standards is not emphasized, with economic, social and human resources being mainly used to improve social and environmental conditions.

Option 4: Growth in economic living standards is not desired, with all economic, social and human resources being used to improve social and environmental living conditions, the aim being for future social and environmental living standards to have increased sufficiently to offset lower economic living standards.

The first choice of the respondents was as follows:
Option 2: 59 per cent
Option 3: 21 per cent
Option 4: 13 per cent
Option 1: 7 per cent.

A similar survey in Australia would not necessarily produce an identical result: here, the concept of 'Development' has been pushed by developers, the media and most politicians (especially in the resource-based states) much more persistently. Nevertheless, it is equally rash to ignore the possibility that Australians might choose a lifestyle which puts more emphasis on human, community and environmental considerations than the conventional wisdom asserts.

In the United States, a 1977 Harris survey asked its respondents to choose a range of options, which resulted in a marked rejection of the concept that 'we have no choice'.[28] The respondents thought that 'finding more inner and personal rewards from the work people do' (64 per cent) was more important than 'increasing the productivity of our workforce' (26 per cent); that 'learning to get our pleasures out of nonmaterial experiences' (75 per cent) was preferable to 'satisfying our needs for more goods and services' (17 per cent); that more emphasis should be put on 'learning to appreciate human values more than material values' (63 per cent) than 'finding ways to create more jobs for producing more goods' (29 per cent); and that 'breaking up big things and getting back to more humanised living' (66 per cent) was better than 'developing bigger and more efficient ways of doing things' (22 per cent).

A Norwegian survey (1979) on the social impact of the economic boom generated by the discovery of North Sea oil indicated that a majority felt that there was an over-emphasis on consumption and not enough emphasis on the psychological, ethical and environmental elements in 'quality of life'.

I propose Jones' Seventh Law:

Every technological change has an equal capacity for the enhancement or degradation of the quality of life, depending on how it is used.

There are many illustrations of this thesis, as outlined in the table below.

Technological change and its effects on the quality of life

Mechanized agriculture	+	Vast increases in output, generally at lower cost
	–	Rural depopulation leads to urban population explosion, high energy use, over-consumption (in rich countries), contributes to diseases of over-indulgence
Mass production	+	Greater availability of cheap goods
	–	Destruction of craftwork, loss of sense of personal involvement (alienation), de-skilling
Motor cars	+	Greater flexibility in transport, wider choice of job and home locations
	–	Urban sprawl, road trauma, pollution, increasing psychological and physical dependence on the car, increasing social costs, destruction of transport alternatives
Drugs	+	Relieve pain, permit major operations
	–	Create psychological dependence and physical destruction
Telephone	+	Provides instant, low-cost universal communication
	–	Provides capacity for universal surveillance, violation of privacy, and control
Aviation	+	Speedy and (usually) safe access to all parts of the world

	–	Instrument for mass destruction in war, increased stress and pace of modern life
Television	+	Instant access to the world, and direct (one-way) contact with the political process
	–	Encouragement of a mass uniform response rather than individual responses: cultural norms set at the lowest common denominator rather than the highest common multiple
Pesticides	+	Save crops and decreases insect-borne disease
	–	Pollute environment (especially waterways), leads to the creation of newer, tough types of pests
Atomic power	+	Thirty-year supply of raw material, will avoid some problems caused by exhaustion of fossil fuels
	–	Increases nuclear hazards by making atomic technology universal and reducing emphasis on development of alternative energy sources
Miniaturization	+	Lower costs, vastly greater capacity
	–	Massive reductions in labour requirements

+ *Enhancement*
– *Degradation*

TECHNOLOGY FOR ITS OWN SAKE?

There are worrying implications when human activities become subordinated to technology so that technique becomes an end in itself. In 1858 Abraham Lincoln and Stephen A. Douglas, rivals for election to the United States Senate from Illinois, engaged in seven public debates before huge crowds in which they discussed the issues which later dominated their 1860 Presidential contest. The technology of reporting was very simple: reporters with notebooks, morse-code operators with telegraph lines, and columns of newspaper type laboriously set by hand. Although the technology was simple, the debates themselves were subtle, complex and profound. In 1980, when Jimmy Carter and Ronald Reagan contested the Presidency, they took part in a single ninety-minute televised programme in which brief policy statements were followed by questions from the press. This programme was instantaneously transmitted in living colour via satellite throughout the world. The technology was infinitely more complex, but the level of debate was degraded in relation to that of Lincoln and Douglas. Technology itself determined the form of the Carter-Reagan debate, with its emphasis on *image*, while content was reduced to triviality and banality to fit comfortably with the expectations of an audience habituated to commercial television.

The medical profession is particularly addicted to high-technology solutions to problems. Australian specialists seem to be conditioned to employ their techniques at the highest level whenever the opportunity arises. Compared to the United Kingdom, Australia has, per capita, twice the number of general surgeons, appendectomies and hysterectomies, three times the varicose-vein strippings, five times as many plastic surgeons and haemorrhoid operations. Tonsils and adenoids appear to be uniquely menacing in Australia, and there is no alternative but to remove them: Sweden has 1·3 tonsillectomies per 1000 people per annum, and the UK has 1·9, compared with 5·35 in New South Wales alone. For every birth by Caesarian section in the Netherlands, Australia has eight.

Professor Bjorn Isaksson, Secretary-General of the International Union of Nutritional Sciences, has estimated that as many as 40 per cent of patients may be starving in technologically advanced hospitals. Basic tasks such as feeding sick people adequately have low priority compared to electronic monitoring, X-rays, ultrasonic imaging and other diagnostic techniques; and the medical reaction to symptoms of malnutrition is to administer intravenous feeding rather than giving the patient appropriate food. Hospital managements, faced with a choice of heavy investment in capital equipment such as scanners, or employing more nurses to provide more intimate personal contact with patients, may well choose the first option, which may not always be to the patient's benefit.[29] The surgeons, physicians and hospital administrators all say: 'But I had no choice . . .'

In cancer research, it has been estimated that more than 90 per cent of available funds is spent on developing new technologies for treatment (e.g. chemotherapy, radiation therapy and genetic mutations), and less than 5 per cent on preventative measures such as changing the social environment. The Williams Royal Commission into Drugs was a notable illustration of forcing technological rather than human solutions on to social problems. The five-volume report provides a cursory examination of the causes of drug addiction, and concentrates on the use of technology in surveillance and control (telephone tapping, opening mail, X-ray devices at airports).[30] It fails to examine technology itself as a possible *cause* of drug dependence through the fragmentation of human experience, loss of faculty, and mounting evidence of frustration and despair in urban society.

Thousands of millions of dollars are invested in developing and providing nuclear energy, with all the attendant risks (including treatment and disposal of wastes, and the serious security and control problems of running nuclear plants) while only tens of millions have been devoted to appropriate solar-energy techniques.[31] What steps can be taken in a democratic society to provide appropriate monitoring of technological change?

MONITORING TECHNOLOGY

Sweden, Norway, Denmark, West Germany, France and the Netherlands have passed legislation requiring consultation between industry and trade unions before labour-displacing technology is introduced.

In Britain, the White Paper on Industrial Democracy proposed that all large companies should be required to give notice to employee organizations before any decisions were made on investment plans, mergers, take-overs, expansion or contraction of organizations, and technological change. (However, the necessary legislation has not been passed.) The Swedish *Joint Regulation in Working Life Act* (1977) gives trade unions and employers a general right to negotiate, but a primary duty is imposed on employers to advise unions before important changes are made at the workplace. The employer also has a duty to provide information about profitability, personnel policies, and other matters of mutual concern. The trade unions are guaranteed three rights: to negotiate on matters of concern, to secure relevant information, and to maintain the status quo until agreement is reached on major issues. The Norwegian Federation of Trade Unions (LO) and the Norwegian Employers Confederation (NAF) have reached a voluntary agreement that unions will be fully informed about proposed technological changes, and that social and environmental considerations will be taken into account before decisions are made. Unions have the right of access to all technical documentation.

The Australian trade-union movement has called for technological monitoring to retard or prevent the spread of some forms of labour-displacing technology. This poses problems which make the control of drugs, tax evasion and organized crime seem ridiculously simple by comparison. The need for such legislation is denied vigorously by those who believe, like the Australian Treasury, that the adoption of technology should be directed by market forces alone; and the mere suggestion of any form of monitoring has been indignantly rejected by technological determinists as evidence of Luddite fanaticism. Australia's problems in this area may be greater than in many other countries because of our position as a junior partner in the English-speaking world. For historical, national, political and cultural reasons, Swedish and Norwegian industry would be unlikely to leave Sweden and Norway even if they felt constricted by regulatory legislation. Australian industry and services, especially when foreign-owned, are far more portable if management opposes technology controls – it may be easier and cheaper to withdraw completely than to conform. It would be inequitable for proposed technology controls to be harder on existing industries than future ones, on Australian rather than internationally owned, on small rather than big ones. It is easy to regulate mining which must be carried out in a particular location – but the extreme portability and intangibility of much service-based industry would make legislative controls as futile as ploughing the sea.

Technological impact statements, similar to the 'environmental impact statements' already enshrined in legislation, must be obtained from industry so that the trade unions, the labour force, parliament, the media and the community at large are aware of what is proposed. Parliament must be given access to independent sources of technological assessment along the lines of the United States Congress Office of Technology Assessment (OTA), which was established in 1973. As its former director, ex-Congressman Emilio Q. Daddario argued, Congress ought to be co-equal with government as regards information on technological issues.

> Congress must be able to ask the right questions. To do this it needs background information and it also needs a better understanding of the secondary and tertiary consequences of technological applications . . . Legislators want to know what were the arguments which entered into the decision making process. How was the final decision reached? What alternatives and options were examined? Or why weren't certain options and alternatives examined?[32]

If parliament is to take any part in the determination of technology's role in our lives, it must flex its collective political muscle and attempt to convert the one-dimensional process of technological determinism into a debate.

Asserting the right to choose

Every nation and every individual ought to assert the right to be adequately informed about what is going on, to be able to make an appropriate choice, or – where considered desirable – to delay making a choice. Those who insist that Australia has no right to choose its own technological forms act 'as if technology has a force and life of its own, independent of political, economic and social forces'.[33] One major argument for the adoption of high technology is that 'If we persevere with obsolete technology, our economy will sink like a stone.' This is, as best, a half truth. The Minister for Industry and Commerce, Sir Phillip Lynch, has estimated that 300 000 Australian workers (i.e. 5 per cent of the labour force) are involved in international competition.[34] In the areas where we are internationally competitive, such as agriculture and mining, we already have the benefit of advanced technology; in the areas where we do not – manufacturing – it is hard to see how we can compete with the ASEAN nations. Even the Myers Report does not see much in this argument.[35] The first people in the firing line when technological displacement occurs are printers, shop assistants, office workers, bus conductors, ticket collectors, bank clerks, insurance salesmen and cleaners: without jobs, without income and without spending power, these displaced people cannot buy the products of the new technology – and our economy may well sink like a stone. In this sense there is a built-in stabilizer in the thrusting new technological era which, in the longer term, will 'contain' it. If a product is unsold and unsaleable, the new technological industries will fail.

It is sometimes argued that Australia should identify 'key industries', especially in their infancy, adopt them nationally and promote them internationally. (Dr Peter Stubbs is an enthusiast for this course.) The strategy is attractive, providing that the right choice is made: there are tremendous advantages in being first (or even second or third) in the production line, so that products from the key industry can penetrate and win worldwide markets. But there is absolutely no advantage to being eighteenth or thirty-second in line. There is a mixture of comedy and pathos in current British attempts to 'get with' the micro-electronics revolution, as Sir Keith Joseph explains to bankers why they should invest in microtechnology, fifteen or twenty years too late. In Australia's case, metals processing and fabrication are the only key industries regularly suggested, apart from uranium-based technologies. This raises important issues about resource planning, especially the huge amounts of energy involved and the cost of providing infrastructures for such highly specialized industries as aluminium smelting (see also Chapter 6).

Vast sums are required for capital investment in heavy industries, but older ones are declining as employers and new ones are based on labour-displacing technology. An investment of $1 million in 1980 would provide for *one* of the following:

1 just over one job in aluminium smelting;
2 a 100-bed old peoples' home;
3 ten restaurants;
4 two medium-budget Australian feature films;
5 one-quarter of the Australian government's annual expenditure on sport and physical fitness;
6 two ALP Federal election campaigns (or one for the Liberal Party);
7 advertising for eight 'Promote Victoria' campaigns by the state government.

All these alternative investments have 'multiplier' effects (like mining) and all are interconnected, yet vehement and well-placed lobby groups (aided by a partly subconscious, partly masochistic Australian yearning for gigantism) effectively prevent any comparative evaluation of employment generation being made.

We must resist the view that we are powerless to defer decisions on particular forms of technology, to have second or third thoughts. Fashions change. A decade ago, supersonic airliners were all the rage and in 1972 Vice-President Spiro Agnew said: 'It must be obvious to anyone with any sense of history . . . that there *will* be SSTs [supersonic transports]. And Super SSTs. And Super-Super SSTs.' Where are they now? (For that matter, where is he?) President Georges Pompidou insisted that 'Paris must adapt to the vehicle', ripped down many old buildings, and attempted to turn his capital into the French Los Angeles – but President Valéry Giscard d'Estaing took a 180-degree turn and said 'The vehicle must adapt to Paris.' Melbourne's trams were marked for extinction in the 1960s to provide more road space for cars – but by 1980 the trams were recognized as a major asset.

We live in an uncertain world; science cannot provide us with certainty. Werner Heisenberg's 'uncertainty principle' argues that in physics we may speak of probabilities but not of certainties – the more minutely we examine anything, the harder it is to be certain about it. Kurt Gödel's 'incompleteness theorem' argues that even in arithmetic it is uncertain that basic axioms will not lead to contradictions. Alfred North Whitehead has drawn attention to 'the fallacy of misplaced concreteness'. Laying down the law peremptorily is not only dangerous, but unscientific as well: we should opt for pluralism and reject the concept of absolute truth. When Australia adopted metrication, *carte blanche* was given to the technologists and parliament abdicated its right to modify legislation so arbitrarily set before it. There were good reasons for adopting the metric system, but regulations setting penalties for people who failed to use metrics gave an ominous insight into the rigidity of the technological mind, with overtones of 'Newspeak' in *1984*. Rulers with Imperial measure on one side and metric on the other were prohibited imports, confiscated by customs officers. Supporting the metric system is one thing, but to impose an eleventh commandment, 'Thou shalt have no system of measurement but metric', is cultural fascism.

The Australian Parliament has proved almost totally incapable of taking any collective position on the social impact of technology, and has retreated into an embarrassed silence. Of Australia's 784 MPs, only one made a submission to the CITCA; there were only two debates on technology in the 31st Commonwealth Parliament (23

October 1979 and 18 September 1980), totalling less than two hours. Politics lives in a crisis atmosphere. Short-term, urgent considerations inevitably receive more attention than long-term, important ones, on the basis that the urgent must be dealt with straight away while the long-term can be examined at some future time, when all the facts are in. In reality, of course, the appropriate time never comes and, as Edward de Bono says, 'The urgent always displaces the important.' Technology develops its own momentum, and can be used as an instrument of the strong against the weak. The only thing that can stop 'the imperialism of instrument reason' is a vigorous revival of the political process and an insistence that changes, both major and minor, be analysed thoroughly and argued out in a spirit of passionate scepticism.

The adoption of technology which will abolish dirty, dangerous and dehumanizing work must be welcomed unequivocally, but we must assert the right to choose appropriate types of technology at our own pace, and to express a preference for those which enhance and extend human capacity, dignity, diversity and understanding.

The marvels of modern technology include the development of a soda can which, when discarded, will last forever – and a $7000 car which, when properly cared for, will rust out in two or three years.[36]

11 What Is To Be Done?

Leisure creates its own demand and to meet it we must build factories for ideas just as there exist factories for machinery.

André Malraux

Making appropriate responses to the changing nature and expectations of work in a post-industrial or post-service society depends, unless we are merely prescribing palliatives, on recognizing and understanding what is happening. Analysis comes first – and ten chapters have been devoted to that. The remedies proposed in this chapter arise from the analysis, and recapitulate my central thesis in a step-by-step programme.

Work opportunities in a post-service era

The greatest hope for future employment opportunities in a post-service era lies in work which is:

1 labour/time-absorbing (and consequently low in productivity);
2 not subject to direct competition from technology;
3 not subject to foreign competition (or needing tariff protection or quotas to survive);
4 not based on large-scale capital-intensive enterprise;
5 not based on producing commodities which have an extended life;
6 not based on a new invention or technological form;
7 low resource-using (not entropic);
8 aimed at the satisfaction of individual needs (e.g. providing a million different garments rather than the same garment a million times);
9 based on fulfilling human needs on a continuing basis (e.g. restaurants, entertainment, sex-related employment), not once and for all;
10 in itself an output of production (i.e. activity for activity's sake, such as professional sport, research, gardening, music and craft, welfare industries).

The most likely areas for future work expansion, not all of them desirable, are in:

1 education, including recurrent education and training for the semi-skilled and unskilled;
2 home-based employment, including domestic work, maintenance and gardening on a contract basis, home security;
3 leisure, tourism, sport and gambling;
4 dining out;
5 provision of drink, drugs and commercial sex (and treating their adverse effects);
6 craftwork, the arts and entertainment generally;
7 individualized social, welfare and counselling services (especially geriatric or psychiatric);
8 individualized transport systems, e.g. taxis, personal drivers, fixed-route minibuses (such as the *peseros* of Mexico), courier services, point-to-point delivery;
9 public-sector employment: administration, armed forces, police;
10 hobby-related work, including DIY work in the informal economy, antiques and collecting;
11 small-unit energy generation (solar, wind, and growing crops for 'biomass'), and subsistence farming;
12 manufacture of leisure and solar-energy equipment (boats, games, solar heaters and collectors);
13 materials recycling;
14 recognizing that some existing forms of work are essentially 'welfare industries' where the main output is *employment*;
15 nature-related work, including gardening in the widest sense: the care and preservation of wildernesses, forests, deserts and natural parks, coastlines, the development and care of footpath networks;
16 care of animals, including selling, breeding and grooming pets.[1]

In the 1970s there was a massive shift in US employment, away from making, selling and servicing goods towards the types of employment listed above – and early warning signs are apparent in Britain, France, Canada, Australia and New Zealand. Between 1970 and 1980, 13 million new jobs were created in the US and 1·5 million jobs in manufacturing were lost – a net gain equal to the entire labour force of Canada. Population shifted away from the traditional manufacturing states of the north and north-east, towards the 'Sunbelt' in the south and west. The new jobs are largely part-time, largely non-unionized, largely for females, largely unskilled – with little prospect of advancement, little job satisfaction and poor job security. (Most could be decimated by a sharp depression.) As Emma Rothschild wrote:

> This transformation has profound consequences for the organisation of society, and for the American economy . . . The United States, in sum, is moving towards a structure of employment ever more dominated by jobs that are badly paid, unchanging and unproductive . . . [for example:] waiting on tables, defrosting frozen hamburgers, rendering 'services to buildings' [including cleaning and maintenance], looking after the old and the ill: 'women's work'.[2]

A political programme for survival and enhanced quality of life

Society is being shaped by a process of political somnambulism in which the parliament is playing a relatively minor role and runs some risk of being hissed off the stage. We feel incapable of taking any positive action to ensure that technology is used to meet social rather than economic goals, or even of assessing if the economic benefits will cause social disbenefits. The following programme will involve reasserting the primacy of the political process:

1 *Changes in economic measurement*

1.1 Recognize that the current definition of the labour force and its conventional three-sector analysis is now inadequate to describe how people are actually occupied, adopt the five-sector analysis set out in Chapter 3, and measure the economic and social significance of work performed in the home. Consideration should also be given to creating a sixth (senary) sector to include students.

1.2 Re-define *work* as 'any form of activity or time use that is or may be beneficial to society and/or to the person performing it'.[3] *Income* should be re-defined to include 'acknowledgment of the right to receive economic support' in addition to 'reward for work done'.

1.3 Add the value of domestic and DIY work to the national accounts in calculating the value of production.

1.4 Adopt the physical quality of life index (PQLI) as developed by the Overseas Development Council, Washington DC, a composite of three indices, each given equal weight: life expectancy, infant mortality and literacy. The PQLI should be expanded to take pollution, drug and alcohol dependence, mental breakdown and suicide rates into account. GDP measurement should be adjusted to allow for dis-economies such as waste, resource depletion, and the cost of pollution control.

1.5 Recognize that the concept of 'high productivity', which usually means 'low employment-absorbing', requires that social costs be taken into account.

1.6 Collect and evaluate statistical data for measurement of the social impact of economic changes, e.g. the significance of technological change in employment, the regional or class effects of differences in educational absorption, the extent of regional and class variations in consumption patterns.[4]

1.7 Consider new methods of calculating and determining income. The concept of the 'social wage' should be examined: this involves deducting the actual costs of transport to and from work, and of access to education and health, from the award wage.

1.8 Computers should be rated in terms of their manpower displacement capacity, in the same way that engines are rated for 'horsepower' (and for similar reasons), to ensure that computers of appropriate capacity are acquired and potential for labour displacement can be calculated. This rating would be measured in GSUs (see Chapter 5).

2 *National labour-force planning*

2.1 Establish a National Labour Force Planning Commission to work with trade unions, employers, states, schools and tertiary educational institutions, local government and Commonwealth authorities to collect and analyse data about changing employment patterns and expectations, assist clients to evolve plans for meeting anticipated social and economic needs at local, state and national levels.

2.2 Evolve national labour-force policies to maximize the range of personal options for employment and assist transitions without trauma.

3 *Education*

3.1 Provide working-class children with the same opportunities and incentives to undertake tertiary and further education as are taken for granted in the upper and middle classes.

3.2 Increase public expenditure on education in disadvantaged areas to the extent needed to offset current social regional inequalities and, where necessary, establish an alternative career structure for teachers.

3.3 Establish the automatic entitlement of every citizen to twelve years of publicly funded education, as recommended by the Australian Poverty Commission.

3.4 Provide differential rewards for teaching services, e.g. higher salaries and accelerated promotion for teachers working in disadvantaged schools, where the retention rates after Year 10 are low.

3.5 Establish a system for two to three years of general education (as in the US college system) before tertiary students proceed to specialist degrees.

3.6 Create an 'open university', along the lines of the British prototype, in which people of all ages can study for degrees at home, in their own time, using television and other modern techniques.

3.7 Encourage the concept of recurrent or lifelong education, so that people are able to 'drop in' or 'drop out' of education at any stage in life.

3.8 Emphasize education as a means of achieving self-knowledge, personal development, the strengthening of self-image and creativity, and effectual time use (including leisure studies), and place less emphasis on education for vocational or specialist purposes which can be picked up relatively quickly.

4 *Income support and superannuation*

4.1 Introduce a guaranteed-income scheme along the broad lines recommended by the Australian Poverty Commission (see Chapter 9).

4.2 Abolish all work tests, which punish the unemployed if they fail to take up inappropriate or remote work by denying them unemployment relief.

4.3 Introduce a national superannuation scheme similar to the Swedish model, with universal cover for those in work and with full portability. (The Swedish system is largely financed by employee and employer contributions – the resulting fund is then invested in industry, generating a large income for beneficiaries.)

5 *Changes in work patterns*

5.1 Legislate to guarantee full and appropriate employment for those able and willing to work.[5]

5.2 Recognize the 35-year working lifetime as the norm, guarantee maximum superannuation benefits at that point, and phase it in *pari passu* to coincide with reduced labour demand. (Superannuation benefits may diminish thereafter – on the grounds that persons remaining in work would aggregate more income but may decrease job opportunities for others. Diminishing benefits may be an economic incentive to get out early, but should not be mandatory.)

5.3 Abolish all mandatory retirement ages. Personal options for early or late retirement should be maximized – workers should not feel under economic pressure to stay on if they would prefer to retire or to retire if they wish to stay on.

5.4 Provide for sabbatical leave for workers on the same basis as enjoyed by some professionals. This would be a major generator of income in tourism, leisure, education and training (many people already seek time off to drop back into education).

5.5 Encourage unions and management to establish options such as a shorter working week, a shorter working year, or a shorter working lifetime.

5.6 Encourage industrial democracy so that workers are involved in decision-making and creative work. The Lucas Aerospace Combine Shop Stewards' Corporate Plan illustrates the productive potential (and social benefits) which could result from encouraging process workers to develop their own skills and knowledge.

6 *Technology control*

6.1 Commonwealth and state parliaments to assert their right to shape, influence and – where necessary – control technological development. If there is constitutional doubt about the power of the Commonwealth to legislate, then this must be sought from the states, and if refused must be put to a referendum so that the issues can be fully discussed with the electorate.

6.2 Adopt an equivalent of the Swedish *Joint Regulation in Working Life Act* to provide that prior information about proposed technological changes should be given by its proponents to appropriate unions, and that negotiations should follow to ensure that any proposed economic benefits are shared by the labour force affected.

6.3 Establish a parliamentary office of technology assessment to provide independent and, where necessary, adversarial advice on technological matters. A joint parliamentary standing committee on technology should also be established.

6.4 Recognize the need to adopt appropriate technology to meet particular needs, rather than accepting the view that technology must either be adopted *in toto* in its most advanced form or, in effect, be totally rejected.

7 *Technological impact statements*

7.1 Commonwealth and state legislation must require the production of technological impact statements which set out:

(a) the nature of the proposed changes;
(b) their anticipated social and economic consequences;
(c) their anticipated impact on employment;
(d) how the relevant decisions were made;
(e) what consultations were carried out with relevent trade unions and government instrumentalities.

8 *Limiting the power of multi-national corporations*

8.1 As the economic and social interests of particular states will not necessarily coincide with the economic aims of multi-national corporations, parliaments should legislate to provide that control of national assets – especially minerals and energy – remain in Australian hands.

8.2 Lending of Australian funds to multi-national corporation operations should be subject to strict control. Australians are lending money to foreign multi-nationals to enable them to 'buy out the farm' – that is, we lose control by allowing our money to be used to finance the loss of Australian equity.

9 *Encouraging small business*

9.1 Government assistance should be provided to secure access to risk capital for small businesses, and to secure expert advice in accountancy, taxation, market analysis and co-operative marketing to assist them to operate effectively.

9.2 Large firms should be encouraged to recognize that the 'hiving off' process – where new forms of enterprise break away from existing entities – has been very successful in the United States and Europe in generating new ideas and new employment. (Many of the most productive enterprises in micro-electronics and biotechnology have been developed by new, small firms and resisted by old, established corporations.) Large companies should also be encouraged to follow the lead of Shell, IBM and British Steel in subcontracting some types of work out to small specialized labour-intensive firms which can absorb redundant workers. This process, encouraged in the USA by entrepreneurial spirit and availability of risk capital, is foreign to the Australian tradition of playing safe, adding new enterprises to existing corporations, and importing both technologies and ideas.

9.3 Municipal authorities should be encouraged to set up 'local enterprise trusts' to contract with young people to engage in domestic repair and maintenance (including retro-fitting for energy conservation), tree planting and beautification schemes.

10 *Information*

10.1 Amend the Freedom of Information Act to follow the US legislation, or at least incorporate the safeguards recommended by the Senate Constitutional and Legal Committee. These include the right of individuals to have access to their own files (and to be able to have errors corrected), the right of access to old files, refusal by a minister to grant access to be subject to challenge in the court, and penalties for bureaucratic delays in providing access.[6]

10.2 Adopt a national information policy, as follows:

(i) All Australians are entitled to free access to information and library services of acceptable standard relative to the affluence of the state – regardless of where they live, their social and economic position, their language, their age, their mobility, or physical disabilities.

(ii) The effective working of democracy depends on the availability of adequate information and the capacity for its independent evaluation. The right to be informed is basic to every person: access to information and the right to its availability should be vigorously pursued. In our society, access to information is a vital resource of both the government and the public, and – like other resources – it should not be concentrated in the hands of the rich.

(iii) The Australian community is divided between the information-rich and the information-poor. Women, migrants, the aged, the young, the poor and the handicapped are victims of inadequacies in the information delivery system. Information facilities are remote from those who need them most: they do not know what is available, and do not know how to remedy their lack of information.

(iv) The enormous inequity in information transfer means that the position of the individual compared to governments and corporations may deteriorate rapidly. The increasing volume of available information may lead to:

(a) increasing tendency towards specialization and the fragmentation of knowledge;

(b) a growing sense of alienation or anomie in many people who feel unable to understand what is going on around them;

(c) a risk that power will move towards the technocrats and away from elected institutions such as parliaments.

(v) Information problems should not continue to be treated in the narrow perspective of science and technology, and left to scientists and experts alone, but should be viewed in the broader context of knowledge and social welfare where information is regarded as a fundamental resource.

(vi) Governments must not allow automated systems and networks to develop in a chaotic fashion for strictly commercial motives. They must understand and plan for the new technologies which will come into general use everywhere in the next ten years.

(vii) As recommended in the Coombs Report[7], a Commonwealth

Information Advisory Council should be set up to define and codify:

(a) the right of access of individuals or public or private bodies to this information resource and the inherent limits to this right;

(b) the political and social guarantees which individuals and institutions can legitimately expect, including protection of privacy and professional secrecy;

(c) the basic rules of reciprocity which should govern relationships between public and private systems and networks;

(d) a code of ethics for professions and industries concerned in this field.

(viii) The provision of public library and information services should remain the collective responsibility of the Commonwealth, state and local levels of government, funded in part by each of these.

(ix) Public libraries must become information resource centres including local data banks and information about access to government services, with greater emphasis on non-book material such as gramophone records, tapes, cassettes, microforms and audiovisual material generally.

(x) The expansion and extension of all library and information services should be achieved through co-operation and/or contract and the formation of library systems and networks, with each service retaining its autonomy within the overall state plan for the development of these services. Already existing information services should be co-ordinated and integrated to avoid duplication and waste of resources.

10.3 Establish an Australian Information Utility as a statutory corporation. A majority of the corporation would be appointed by the government of the day, with minority representation from business interests (who would presumably be the major users) and community groups – trade unions, political parties, consumer organizations. The government members would include representatives of Telecom, the Commonwealth Scientific and Industrial Research Organisation (CSIRO), the ABC, the major libraries, and at least one Opposition member of parliament.

Initially terminals for public use would be provided in accessible locations such as schools and libraries, but after Telecom adopts digital transmission it will be possible for home telephones to be part of the international telematic system. The utility would control the Australian communications satellite. There are a number of technical problems to be overcome – the question of protecting property rights in copyright material, maintaining privacy and confidentiality, and calculating appropriate tariffs. The system would be open for all at cheap rates. It could be a valuable instrument to aid the weak against the strong – which is the main reason why we are unlikely to adopt it.

10.4 Stimulate public debate and information flow by establishing two bodies based on the New Zealand 1977 Planning Act:

(i) A planning council to examine short-term implications of technologically based social change.

(ii) A Commission for the Future to examine medium- and long-term implications of technologically based social change.

11 *Arts, leisure, craftwork and sport*

11.1 Increase expenditure to stimulate arts, leisure, craftwork and sport, both for their intrinsic worth in promoting quality and diversity of life, and to promote personal development, physical and mental health, and reduce socially costly dependence on drugs, welfare and health care. These areas have considerable employment potential.

12 *Job creation*

12.1 In the short term, it will be necessary to set up job creation schemes to provide work opportunities, particularly for young people, in regions suffering from heavy unemployment. Being out of work for six months or more is extraordinarily destructive to the self-confidence of young people, confirms their low self-assessment, and leads to major social traumas including apathy, drug abuse, alcoholism, crime and suicide. The restoration of self-confidence is the primary aim of job creation – its contribution to GDP will be minimal. In the longer term, job creation will be merely a palliative: forms of boondoogling or 'Potemkin village' work. However, there are significant opportunities to create service employment in areas which have an atypically high manufacturing labour force. Creating new jobs in manufacturing is very expensive, often more than $200 000 in capital for each one. In service employment such as restaurants, shops, beauty care and sports facilities, the cost of job creation is far lower (often less than $10 000 per job).

13 *Domestic work*

13.1 As interim measures in recognizing the economic value of work performed within the home, provision should be made for workers' compensation benefits, and for increased home-relief services (to give men and women free time for shopping, study or other personal needs, such as holidays).

13.2 People undertaking care of the aged at home should receive not less than the government subsidy to old peoples' homes for each inmate.

14 *Research and development*

14.1 Increase expenditure on research and development in Australia. Australia ranked thirteenth of seventeen OECD nations in expenditure on industrial research and development, largely because of the low level of business investment. The Senate report *Industrial Research and Development in Australia* criticized the 'dismal decline in public and private research and development'. Australia, which has given the world the Nomad, Interscan, the Hills Hoist, Vegemite, the lamington and the Chiko Chicken Roll, should have far more to offer. An Australian computer industry should be supported, although it will not be a major employer.

14.2 Investigate 'appropriate and light-capital technology' (see Chapter 10).

15 *Decentralization: alternative lifestyle*

15.1 More employment may be generated by reversing the economies of scale so that fewer things are produced by more people. While this does not satisfy the test of economic cost-effectiveness, it makes good social sense. Decentralization is an option which needs investigation – although Australia's strong urban tradition suggests that its scale will be limited. Guaranteed income may encourage more people to leave the urban labour force and pursue alternative lifestyles largely outside the conventional economy.

16 *Trade training*

16.1 Employers who do not train their own workers should pay a levy to contribute to the costs of trade training.

16.2 Conventional apprenticeship, which teaches a narrow range of skills, should be phased out and replaced by 'modular' or 'polyvalent' training in four or five diverse (but related) skills, to provide greater flexibility for future employment needs.

17 *Taxation*

17.1 Australia's taxation system, which raises the cost of employing labour (by imposing a payroll tax) and reduces the cost of capital equipment (by providing generous investment allowances), should introduce some degree of fiscal neutrality by reducing the cost of labour and increasing the cost of capital.

17.2 Payroll tax, which is an economic disincentive to employment, should be replaced by a levy on turnover. (Payroll tax is remitted to the states as a growth tax: in 1979–80 this raised $1685 million, and its elimination would be a serious blow to the states unless replaced by some alternative revenue.)

17.3 Taxation policy should distinguish between investment allowances for capital equipment which produces a net growth of jobs, and capital investment which destroys jobs.

17.4 A 'human capital' allowance should be provided for, so that the costs of training and retraining can be written off or depreciated like a capital investment in plant equipment.

18 *Union consolidation*

18.1 Australia has 280 trade unions, of which half (with only 6 per cent of total union membership) operate within a single state. West Germany has only seventeen unions: this has prevented union strength being dissipated through demarcation disputes, and enables unions to maximize and co-ordinate their research and negotiation capacity to protect the long-term interests of members. Consideration will have to be given to consolidating existing union structures in Australia.

19 *Strengthening the international trade-union movement*

19.1 Australia should push, through the ILO, for the strengthening of the international trade-union movement to ensure the universal right of unionists to elect their own officials, gain legal protection for their activities, and be able to negotiate with governments and employers.

19.2 Since multi-national corporations operate across political frontiers throughout the world, trade unions must be able to operate as an international countervailing force, and to collect and share available information.

19.3 In ASEAN and other low-wage areas, it is essential that trade unions be strengthened so that wage levels in particular industries move towards parity in order to avoid dumping, industry instability, trade wars, and moving offshore.

20 *Sharing social benefits and costs of technological change*

20.1 Budgetary strategies should ensure that gains from technological innovation are shared to maximize community benefit, and to avoid benefits being confined to owners of technology and disbenefits to displaced workers.

20.2 The social cost of providing for job displacement must be the primary responsibility of those deriving economic benefits from the adoption of job-displacing technologies: the first charge should be set against profits, the second against general tax revenue.

21 *Adjusting balances between factors of production*

21.1 Government can, without undue market distortion, adjust the balance between different factors of production. Examples include providing financial disincentives for energy- and capital-intensive activity, and incentives for labour-intensive activity – such as discouraging high fuel usage and encouraging energy conservation and retro-fitting – or emphasizing the need for superior-quality goods and after-sales service rather than the packaging, promotion and advertising of low-quality high-volume goods.

22 *Prices and incomes policy*

22.1 It will be necessary to negotiate national agreements on prices and incomes as an anti-inflationary measure in a time of rapid transitions in employment, largely for the psychological impact it will have on inflationary expectations.

RAISING LEVELS OF CONSCIOUSNESS

Legislating for industrial democracy, providing new means of re-distributing wealth, enabling people to have greater freedom to move in and out of the labour force, and extending access to education, culture and other forms of self-expression are only part of tackling the problems of structural and personal adjustment in a post-service society. The most difficult changes are essentially conceptual and relate to people's image of themselves, their place in the world, their goals and capacities. In past eras, massive changes such as the steam engine and electrification occurred gradually over decades and it was possible to make adjustments (painful as they were in many cases) over a lifetime. Now changes occur very rapidly, and will continue to do so, but human adjustment time is as slow as ever with the prospect of several major dislocations in each decade. We will, individually and collectively, need to raise our levels of consciousness if we are to get the best and not the worst out of the coming changes – to play active, life-enhancing roles rather than negative, defensive ones. These changed levels will cost nothing to implement, and will make a major contribution to mental health. They cannot be legislated for – we must accept or reject them individually. Important changes of heart and mind are needed:

- We need to think of work as any kind of worthwhile activity, whether paid or not, and to recognize housewives, students and hobbyists as being usefully employed.

- We should recognize income as including a right to receive economic support, not just as reward for work performed.
- We should guarantee the right to work for those willing and able to do so, regardless of age. We should also decide that the right to work must also include the right *not* to work and that the compulsory work ethic is now irrelevant, obsolete and counterproductive. Many people may feel a temporary sense of moral outrage if work ceases to be compulsory, but it will not lead to social disintegration.
- We should realize that technology has been far more influential than ideology, elections, political struggles, or education in changing the way people live. Technology, while neutral or 'value-free' in itself, in the hands of its owners or controllers becomes a political instrument for reshaping society, and this power is exercised to a degree that even totalitarian governments would hesitate to attempt.
- We should remember that all knowledge is inexact, and forecasting is difficult. Economics is notoriously uncertain, and past performance by its practitioners does not inspire confidence in their Panglossian optimism. Technology may provide answers for all human problems, but we should beware of making too large a psychic investment in technology and not enough in human capacity. We need moods of balance, open-mindedness and humane common sense.
- We should promote scepticism and reduce credulity. There are too many people who accept everything they are told – especially if they are told the same things often and loudly enough – and too few honest sceptics. Many powerless people feel instinctively repelled by the complexities and seeming omnipotence of science: they turn not to scepticism, however, but to alternative forms of credulity – the occult, religious cults, the drug culture, astrology and prophecy.[8]
- We should apply more flexible intellectual approaches to problem-solving – i.e. lateral as well as linear thinking.
- We should oppose excessive reliance on 'reductionism', which poses grave dangers to democracy and personal autonomy.[9]
- We should reject technological determinism, and restore the significance of both parliament and the political process. Democracy has its failures, as the past successes of Barabbas, Hitler and Peron indicate, but legislatures – although imperfect – have a greater sensitivity to human needs than any other political instrument. If we do not revive the authority of democratic practice we

will drift into a technocratic authoritarianism. Parliamentarians must raise their consciousness and levels of debate on technology, tapping alternative sources of advice.

- We should recognize the need to share (and recycle) resources rather than compete for them, and attempt to apply limits to forward throughput (not limits to *growth*).
- We should recognize that over-emphasis on 'economic man' has led to the eclipse of ethics in policy-making. Human beings need psychic income too, and meeting physical needs alone leads to inner starvation. (The work ethic is essentially economic and materialistic, not a branch of the moral law.)
- We should assert the right to choose between varieties of technology. The futures debate in Australia, such as it is, has encouraged the simplistic view that technology must either be adopted *in toto* in its most advanced form, the sooner the better, or rejected *in toto*, by adopting Buddhist economics, foresaking the flush toilet, the electric blanket and colour television. The debate has ignored our right to select appropriate technologies and to fit them to social needs. 'Technology' is not a monolith – there are multitudes of diverse technologies.
- We should recognize that labour input has a declining relevance to levels of output in technologically advanced societies. The actual hours worked and the numbers of people employed have a direct relationship to the output and profitability of small-scale industry and distribution, but are of diminishing relevance to large-scale, internationally competitive industry.
- We should grasp that the basic values involved in the feminist and ecology movements are life-enhancing and contribute significantly to quality of life in the long term.
- We should adopt different measurements for social and economic well-being to those currently in use.
- We should recognize the central role of education (broadly defined) as an instrument of personal development and extend its benefits to the working class.
- We should avoid over-dependence on any technological form, whether electronic or pharmaceutical.
- We should recognize the social, political and economic implications of unequal access to information. In a computer-based information society, the gap between the information-rich and the information-poor is not narrowing (which is technologically possible), it is widening.
- We must recognize that the elimination of much routine, repeti-

tive, boring and/or physically arduous work, no matter how desirable from academic, aesthetic, bureaucratic or entrepreneurial viewpoints, represents a massive threat to the autonomy of people whose lives have been built around such employment and who lack the requisite education, cultural conditioning, temperament or personal capacity to make a transition to more stimulating or rewarding work. Reticulated water supplies have been an unqualified boon – but not to the retrenched water-carriers. The question of what is to be done to assist the unskilled and semi-skilled to find meaningful and satisfying occupation should be a dominant issue in the 1980s. New work will evolve in the long term – but the lags may cause intolerable social trauma.

- Finally, we need to grasp that the uniformity and regularity of technology in general, and computerized transactions in particular, lead to excessive standardization which underrates or ignores personal factors and variations in individual intelligence. As the computer says in Jean-Luc Godard's film *Alphaville* (1965): 'Reality is too complex for oral expression: it has to be converted into another form.' In the 1980s we must look for individual solutions to human problems, not mass ones.

What are our prospects?

We face an extraordinarily ambiguous future. Technology can be used to promote greater economic equity, more freedom of choice, and participatory democracy. Conversely, it can be used to intensify the worst aspects of a competitive society, to widen the gap between rich and poor, to make democratic goals irrelevant, and institute a technocracy. We must evolve policies in response to the current era of rapid technological change. However, we must first attempt to understand what is going on.

There is nothing inherently alarming about much of the new technology (although artificial intelligence and cyberveillance appear to be exceptions), but there is much to worry about in terms of human responses to it (or, even worse, failure to respond at all). Worst of all is a fatalist acceptance of technological determinism – the belief that nothing can be done to moderate or monitor the social impact of technology. The Western world is passing through a period of technological change, more far-reaching – and much faster – than at any other time. This rapidity and intensity presents massive challenges. The technicalities are easy to solve; the human problems – understanding complex and shifting technological matters and assessing their social impact – are more difficult. Many people feel reluctant to face up to the issues and would prefer to put off thinking about them in the hope that either the problems will go away or prove to be exaggerated, or someone else might solve them (or take the blame if they propose solutions which then prove to be wrong – or unpopular). Every major technological change alters the way people live. The Industrial Revolution caused decades of misery, suffering, exploitation and unemployment to people uprooted from their rural environment. In the long term, standards of living have risen enormously and the needs of an expanding population have been met due to changes in production – but transitional dislocations, unemployment and trauma have often remained for decades. Technology *always* produces social change.

Australia is moving towards a 'corporate state' in which major areas of society are run autonomously: e.g. industrial relations are left to the employers, unions and the Conciliation and Arbitration Commission, and decisions about hours of work would not be made by parliament. Ministers are increasingly dominated by their departments – which favour narrow, administrative resolutions of problems instead of policy decisions based on value systems. International trading matters are left to the market; multi-national corporations and media monopolies are too powerful to touch. The areas parliament can tackle are increasingly limited. Is Australia still a democracy in the traditional sense, when so many vital subjects have been de-politicized? Will technology be in the hands of business? Government? Community groups? Will political decisions be taken, or will they be resolved by 'natural selection', without any political debate? Will Australia have the intelligence, energy or guts to impose democratic and pluralist forms on the new technology, or will its ambiguities all be resolved in favour of the rich, the powerful and the status quo? Our timorous social history, the feeble grasp of complex matters exhibited by too many of our leaders, the low level of intellectual vitality, a lack of national self-confidence, our natural tendency towards bureaucracy, conformity, obedience and fatalism, the mediocrity of the business and academic establishment do not give us much ground for optimism.

The impact of technology will be experienced both in the microcosm – the short-term effects in Australia – and in the larger, historic world context. If changes in the pattern of work lead to people being compulsorily retired at fifty-five or unemployable at twenty-five, we should not be surprised if they turn to liquor, drugs, daytime television, the occult, introspection, boredom or emotional paralysis. We must not waste our greatest national resource – people. If we have an alienated segment of young people permanently excluded from the labour force, we should not be surprised to face urban terrorism along Baader-Meinhof lines in Australia before we are far into the 1980s. In the larger context, matters are even more serious. Machines are doubling their intellectual capacity every few years, but people are not. If artificial intelligence outstrips human intelligence, if technology is smarter than its displaced human equivalent, then the power of the people who own the machines will be expanded to an almost unimaginable degree. What are the implications for our political system? In Australia, the current generation of managers grew up before the technological revolution. They do not fully understand its significance – and have an instinctive anxiety that if the number of Indians is reduced, fewer chiefs will be needed as well. The sheer incompetence of Australia's current management is for the time being an asset in maintaining relatively high employment levels. But we cannot count on that incompetence for ever. When the existing technology is used at full capacity, or when new generations of managers arrive on the scene, the impact may be enormous unless we adopt appropriate social responses. It is time to examine the implications.

The fragile consensus which links the Australian community can be shattered if we fail to grasp the interdependence of the skilled and unskilled, rich and poor, market sector and convivial sector. It is essential to recognize the need for employers, trade unions, major political parties and all levels of government to evolve broad policies to ensure that technological change is not used to widen social and economic divisions, and avoid a legacy of increasing bitterness between the powerful and the impotent.

The crisis consists precisely in the fact that the old is dying and the new cannot be born. In the interregnum a great variety of morbid symptoms appear.

Antonio Gramsci

Appendix: Jones' Seven Laws

1 Technological innovation tends to reduce aggregate employment in the large-scale production of goods and services, relative to total market size, after reaching maturation (e.g. transition from manual telephone systems which were labour-complementing to automatic systems which were labour-displacing) and to increase employment at lower wage rates in areas complementary to those technologically affected. (Chapter 2, p. 44)

2 Employment absorption tends to be in an inverse relation to economic efficiency, to a chaos point of inefficiency beyond which labour is not absorbable. (Chapter 4, p. 93)

3 In the production of goods or market services on a massive scale, employment tends to be in inverse proportion to demand. (Chapter 6, p. 139)

4 The economic viability of a technologically advanced society depends on having an increasing number of small consumers despite a contracting number of large producers. (Chapter 6, p. 143)

5 Rising levels of employment depend on increased demands for a diversity of services, many stimulated by education: simplicity of personal needs contributes to low levels of employment, and complexity to high levels. Over-specialization and economic dependence in particular regions on a single employment base (e.g. heavy industry or farming) inhibits the development of service activity. (Chapter 6, p. 148)

6 The amount of time spent by generalists in making technically based decisions is in inverse proportion to the complexity of the subject matter. (Chapter 8, p. 176)

7 Every technological change has an equal capacity for the enhancement and degradation of life, depending on how it is used. (Chapter 10, p. 231)

Notes

Full details of references cited here can be found in the bibliography on p. 271.

1 FROM A PRE-INDUSTRIAL TO A POST-SERVICE SOCIETY

1 The 1780s was marked by significant change in many fields: the US Constitution (1788) and Washington's inauguration as first President (1789); the beginning of the French Revolution (1789); Joseph II's attempt to convert the Holy Roman Empire into a unitary state based on rational principles; the high point of Enlightened Despotism (Frederick the Great and Catherine the Great); European settlement of the last available continent (1788); the first manned international flight (France to England – 1785); James Watt's double-action steam engine (1782); the first textile factory run by steam (1785); the first steam boats; Mozart's greatest operas, concerti and three last symphonies; determining the composition of air and water by Henry Cavendish and others, identification of uranium, tungsten, zirconium; discovery of the planet Uranus, James Hutton's 'uniformitarian' theories in geology, and founding the London *Times*.

2 In 1979, 70·5 per cent of Australia's paid labour force was engaged in services if 'building and construction' (7 per cent) were included. If they were included in the secondary sector (as I have done in ch. 3), then the services total falls to 63·5 per cent.

3 *Public Administration Review*, May–June 1971.

4 Reprinted in *Essays in Persuasion* (Macmillan, 1931).

2 AN AGE OF DISCONTINUITY

1 The story of Cain and Abel may well reflect Neolithic Revolutionary tension between farmers and shepherds. Tubal-cain, six generations on, was 'instructor of every artificer in brass and iron' (Genesis, iv, 2; 22).

2 Many inventions were worked on independently and simultaneously, e.g. steam boats (Rumsey, Fitch, Stevens, Fulton, Symington), the electric telegraph (Wheatstone, Morse), the incandescent light-bulb (Edison, Swan), and motion pictures (Muybridge, Edison, Lumière).

3 Reprinted in American Economic Association, *Readings in Business Cycle Theory*, 1950. Kondratiev was arrested for political crimes in 1930 and sent to Siberia. He has not been rehabilitated. The *Great Soviet Encyclopedia* dismisses his work as a 'vulgar bourgeois theory' which attempts to defend capitalism, and does not refer to his fate.

4 This view is also taken by Geoffrey Blainey in *The Causes of War*.

5 Technological change often accompanied high population growth rates (e.g. the UK, US and Japan), but there have been many exceptions: Ireland, China, Egypt and North Africa had rapid increases in population between 1801 and 1850 with little technological change. In the 1970s, high technology was a feature of nations with *low* population growth.

6 Statistical material on national education since 1780 is extremely scanty. The chart on p. 14 contains rough estimates of proportions of children undergoing systematic education between the ages of five and seventeen in England and Wales. Literacy rates seem to have been high (but the standards were modest): about 50 per cent in 1780, 67 per cent by 1850, 97 per cent by 1900, and about 98 per cent at present.

7 It is estimated that there were 80 000 prostitutes in London in 1851. In 1911 there were 2·6 million domestic servants in Great Britain. Even Karl Marx had one servant and sometimes two.

8 E. P. Thompson, *The Making of the English Working Class*, pp. 295–6.

9 See Malcolm I. Thomis, *The Luddites: Machine Breaking in Regency England.*

10 E. J. Hobsbawm, *Industry and Empire*, p. 94.

11 Lewis Mumford, *The Pentagon of Power*, p. 153. Mumford, a disciple of Sir Patrick Geddes, has written on architecture, history, philosophy and social criticism.

12 Smith (1723–90) was Professor of Moral Philosophy at Glasgow University. His circle included Dr Johnson, David Hume, Edward Gibbon and James Watt.

13 Karl Marx, *Capital* (tr. Ben Fowkes), vol. I, ch. 25.

14 Milton Friedman was Professor of Economics at the University of Chicago 1946–77, and won the Nobel Prize for Economic Science in 1976.

15 Britain was engaged in the following major wars: against the United States 1780–83, 1812–14; against France 1793–1802, 1803–14, 1815; against Russia 1854–56; against South Africa 1899–1902; against Germany 1914–18; against Germany and Japan 1939–45. I have not counted the Indian Mutiny, a long series of colonial wars and the Korean War, where comparatively small forces were committed.

16 See Raymond Williams, *Keywords*. Under 'unemployment', Williams cites E. P. Thompson for the 1820s to the 1830s, but notes G. M. Young's view that 'unemployment' is first used in the 1860s. Milton (1667) used 'unimploid' as meaning idle.

17 Karl Marx's *bête noire* Andrew Ure, the panegyrist of the First Industrial Revolution who argued that working in gas-lit factories was as healthy as working in open fields, was a contemporary of Carlyle.

18 *Technological Change in Australia*, Report of the Committee of Inquiry into Technological Change in Australia (Chairman: Sir Rupert Myers). 'Total cost to the Commonwealth [was] . . . unknown', but expenditure against the appropriation for CITCA was $514 038 (see *Hansard*, House of Representatives, 5 December 1980, p. 468).

19 The curious use of the word 'created' suggests that the committee considered technological change in complete isolation from historical factors. The preface to vol. 2 of the report (also curiously) states: 'Much of the information that would be necessary to enable such assessments [of technological unemployment] to be made cannot be collected because it relates to events that have not yet occurred' . . .

20 *Hansard*, House of Representatives, 18 September 1980, p. 1518.

21 *The Implications of Technological Change*, by Ieuan Mapperson, commissioned and published by the Committee for Economic Development of Australia in 1979, also puts heavy emphasis on television as an example of technologically induced employment growth. The fine flow of his argument is not impeded by the use of statistics. The paper was praised in the *Australian Financial Review.*

22 Yoneji Masuda, *Social Impact of Computerization: An Application of the Pattern Model for Industrial Society.*

23 As late as 1818, Thomas Martin's *Cyclopedia of the Mechanical Arts* (London) failed to mention Watt's steam engine. It would occasion some comment if a 1982 encyclopedia failed to mention electronic computers.

24 George Steiner has observed that when Winston Churchill first went to war in 1899 the fighting would have been comprehensible to Homer. Churchill died in a thermonuclear age.

25 Since the 1970s, the classical 'equilibrium' theory of employment has looked increasingly dubious. A hydraulic analogy may make help to make the point: given two water tanks – one marked S (supply) and the other D (demand) – joined by pipes, as the level in one tank fell it would alter the level in the other. But if the S tank is located in Nation A and the D tank in Nation B, then equilibrium will be of

declining relevance within the individual nations (although it may continue to have global significance).

26 The *New Statesman* once ran a competition for an entirely original sin. The results were extremely disappointing variations on old ones.

27 Between February 1974 and February 1979, *Futures* (IPC, London) ran a valuable series of twenty articles about past technological predictions. They should dispel the illusion that new industries arise without giving early warning signals.

28 Dr Peter Stubbs, in *Technology and Australia's Future: Industry and International Competitiveness*, is enthusiastic about air conditioning in motor vehicles.

29 Most of the winners of the Nobel Prize for Literature in the period 1970–80 were unknown outside their homelands (even inside in some cases) – Solzhenitsyn, Böll, White, Bellow and Singer notwithstanding.

30 Ellul was a Resistance worker during World War II, Professor of Law and Social History at Bordeaux University, and the author of many books. His work is discussed further in ch. 10.

3 A NEW ANALYSIS OF THE LABOUR FORCE

1 Fisher (1895–1976) was educated at Melbourne University and the London School of Economics, held Chairs in Otago, Perth and London, and later worked for the International Monetary Fund.

2 Clark was educated at Oxford, and divided his long career as an economist between Britain and Australia (especially Queensland).

3 Printing is becoming increasingly remote from its historic industrial base, and the linotype is virtually a museum piece. The printer is now more closely related to operators of typewriters, telexes, word processors and duplicators than to process workers in factories. Printing is essential to the legislative process, but the existence of a printing shop does not turn Parliament House into a factory.

4 The absolute numbers of information workers increased from 37 million in 1970 to 44·6 million in 1980, but the total labour force increased by nearly 16 million. See Daniel Bell, 'The Information Society' in Michael L. Dertouzos & Joel Moses (eds), *The Computer Age*.

5 This proposal was first set out in the House of Representatives in February 1978 (*Hansard*, vol. 108, pp. 164–8), further elaborated in later speeches, in a paper 'The Challenge of Post Industrialism' (July 1978), and in 'Implications of a Post Industrial or Post Service Revolution on the Nature of Work', a submission to CITCA in July 1979.

6 See Scott Burns, *Home Inc.*

7 A. Madison, 'Long-run Dynamics of Productivity Growth' (1979), quoted in Professor H. Lydall, 'Technological Change and Economic Growth', in Myers Report, vol. 4.

8 See Graeme Davison, *The Rise and Fall of Marvellous Melbourne*.

9 Hayter's Victorian figures were undoubtedly superior to those of New South Wales, but they do contain puzzling discrepancies. In tables for 1881, the total number of people listed in various manufacturing occupations is 77 841 (more than 20·5 per cent of the paid labour force), but elsewhere in the 1883–84 *Victorian Year Book* he writes of employment in 'manufacturies' as having 'risen' to 46 857 (about 12 per cent). The larger figure may include carters, sales agents and some types of casual labour. I have redistributed the large category of 'labourers (branch undefined)' pro rata to the size of the other sectors, excluding 'information'. Hayter won an international reputation as a statist and was awarded a CMG, and honours from France, Italy and Japan.

10 An earlier version of my four-sector graph appears in the Myers Report, vol. 2, p. 22, and in Henry Mayer & Helen Nelson (eds) *Australian Politics: A Fifth Reader*, p. 65.

11 'Some Implications of the Growth of the Mineral Sector', in *The Australian Journal of Agricultural Economics*, vol. 20, no. 2, August 1976. The 'Gregory thesis' was confirmed by the dramatic *de facto* revaluation of British currency in 1980 – a time of high unemployment and inflation – as a result of North Sea oil discoveries.

12 Nevertheless, preferential purchasing (e.g. where the Victorian government buys Victorian goods at a higher price than goods available from other states) constitutes a *de facto* internal tariff.
13 With printing excluded, the 1947 figure would be 24·06 per cent.
14 And some others (e.g. Frank Hainsworth in his textbook *Economics*). The Hoover Commission in the United States (1945) made the same point.
15 Hugh Stretton, *Capitalism, Socialism and the Environment*, pp. 183–4. Stretton was Professor of History at Adelaide University, demoting himself to Reader so that he could devote himself to applying historical method and lateral thinking to social problems.
16 Figures elsewhere are: US (1970) 4 per cent, UK (1970) 2 per cent, and France (1967) 18 per cent. See Fred Hirsch, *Social Limits to Growth*, ch. 3.
17 Emma Rothschild, 'Reagan and the Real America', in *New York Review of Books*, 5 February 1981, citing US Department of Labor Statistics, *Employment and Earnings*, 1980.

4 TWO TYPES OF EMPLOYMENT AND TIME USE

1 Professor Jean Fourastié examined this concept in his book *Le grand espoir du XXe siècle* (1949). My attention was drawn to his work after I had written this chapter.
2 *Capital*, vol. I, p. 534. This is discussed briefly in P. D. Anthony, *The Ideology of Work*, but I have found no other reference.
3 It was not Smith but his contemporary Benjamin Franklin who stated the adage 'Time is money' (in *Poor Richard's Almanack*), but he would have approved the sentiment.
4 The term 'division of labour' is really a misnomer. A division of labour from the employer's point of view is, to the worker, 'unification' or 'consolidation'.
5 Marx quoted this passage in *Capital*, vol. 1.
6 Harry Braverman, *Labour and Monopoly Capital*, pp. 81–3.
7 ibid., p. 86
8 V. I. Lenin, *Selected Works*, vol. 2, pp. 663, 694. The irony of Lenin's advocacy of Taylorism is reflected in the old joke: *Q.* What is capitalism, Comrade? *A.* The exploitation of man by man. *Q.* And what is communism, Comrade? *A.* Just the reverse.
9 ibid., vol. 3, p. 475.
10 In 1939, when Britain still ruled India's 400 million people, there were only 1007 members of the Indian civil service (excluding base-grade clerks).
11 Dennis Gabor, *Inventing the Future*, pp. 19–20. I thought the dichotomy between Parkinsonian and Smithian employment was mine: I had read Gabor's book nearly twenty years ago, and obviously his message had stuck. A Hungarian, and Nobel Prizewinner in Physics, Gabor worked for many years as an engineer in industry before turning to research and academic work in Britain. Professor Parkinson was not aware of Gabor's work.
12 Bertrand Russell, *In Praise of Idleness and Other Essays*, p. 13.
13 Jack Burnham, *Beyond Modern Sculpture*, p. 11.
14 Ivan Illich, *Energy and Equity*, pp. 30–1.
15 Michael Young and Peter Willmott (op. cit.) found that this was not true of London, which has excellent public transport. They concluded that commuting time had been remarkably stable for centuries: Londoners tolerate about thirty minutes of travelling time each way every day, but use developments in transport technology to provide themselves with additional living space. Ronald Higgins, in *The Seventh Enemy*, suggests that a worker may use more skill and judgment driving to and from work each day than in doing his job – so that private transport has a psychological justification quite apart from its economic value.
16 Ivan Illich, *Energy and Equity*, p. 30.
17 Permanent appropriations, which cover social security, veterans' affairs, and payments to the states, do not come before Parliament for review.
18 In June 1978, California's voters approved (by a margin of 65 per cent to 35 per

cent) a referendum proposition which imposed an upper limit of 1 per cent on taxes on real property and required two-thirds majorities for the imposition of new taxes either by the legislative or referendum. Propositions setting upper limits to public expenditure were carried in thirteen more states in 1978 and rejected in four others.

19 Letter to the author, 12 February 1979. Maddock was Secretary of the British Association for the Advancement of Science, and is now Principal of St Edmund's Hall, Oxford.

20 Hazel Henderson, *Creating Alternative Futures*, pp. 388–9.

5 COMPUTERS AND EMPLOYMENT

1 Paraphrased from Anthony Chandor, John Graham & Robin Williamson *A Dictionary of Computers*.

2 There are some resemblances between the operation of computers and human brains. The dominant hemisphere of the brain (the left side with typical right-handed people) is – like a digital computer – linear, time-orientated, analytical, mathematical and reductionist, better in dealing with specifics that with generalities. The non-dominant hemisphere is – like an analogue computer – holistic, and can grasp complex spatial relationships, respond to feelings (e.g. communicating with animals, young children and people lacking a common language), colours, sounds, movement and simultaneous happenings.

3 Binary notation is very ancient, almost certainly older than the decimal system (originated by the Babylonians, the Hindus contributing the zero) or the quinary system (based on multiples of five) which evolved in South America. In binary notation, adding *0* to the right-hand side of the digit *1* indicates a multiplication by *two* so that, for example, a binary *10* equals decimal *2*, and so on.

4 Alan Mathison Turing (1912–54) was an English prodigy who studied at King's College (Cambridge) and Princeton. His 1936 paper was called 'On Computable Numbers, with an Application to the *Entscheidungsproblem* [halting problem]'. He worked on cryptographic analysis during World War II, built a large decipherer called 'Colossus' (1943), and worked on British computer projects – ACE and MADAM – as Reader in Mathematics at Manchester. Turing was in the tradition of great British eccentrics. He died at 42, probably by suicide. Alonzo Church, born in 1903 and a specialist in symbolic logic, held Chairs at Princeton and UCLA.

5 John von Neumann (1903–57), a Hungarian prodigy, migrated to the US in 1930 and wrote 150 papers on quantum theory, pure mathematics, logic, meteorology, games theory and computer programming (including 'random' factors to approximate human decision-making). He held a Chair at the Institute of Advanced Study in Princeton, New Jersey, and wrote *Can we Survive Technology?* in 1956.

6 Few of us demand information with quite this degree of urgency. However, there are many computational problems which are far beyond the capacity of the fastest available computers – for example, to exhaustively calculate all possible outcomes of an early move in a chess game, using the fastest available computers, would require more than the estimated age of the universe.

7 Wiener (1894–1964), was a US mathematician who won a Harvard PhD at the age of eighteen, and later taught at MIT. He wrote *Cybernetics: or, Control and Communication in the Animal and the Machine* in 1948.

8 *Technology and the American Economy*, Report of the National Commission on Technology, Automation & Economic Progress (1966), p. 1. The commission included Daniel Bell and Robert M. Solow of MIT.

9 This report was enthusiastically endorsed by the *Australian Financial Review*, and has been quoted by directors of IBM Australia in speeches and papers. The report's authors were required to produce an optimistic document at short notice, at a time when an election seemed imminent: they disclaim personal support for the views expressed, and greeted news of the above attitudes with incredulity and hilarity.

10 *Hansard*, House of Representatives, 17 September 1980, p. 1455. The answer confirmed the findings of Peter MacGregor of P. K. MacGregor & Associates

(computer systems analysts of Melbourne) that a data-base search in 1979 of 72 526 English-language citations of sociological abstracts in the United States failed to locate any which had in its title a conjunction of the words Computer/Computers/ Employment/Unemployment.

11 Simon Nora served in the Maquis during World War II, graduated from the *Ecole Nationale d'Administration*, and worked for the French government, the EEC and the publishing industry. He became *Inspecteur-Général des Finances* in 1974. Alain Minc was an *Inspecteur des Finances.*

12 *The Computerisation of Society*, pp. 34–7.

13 I adopt the language of Fred Hirsch (op. cit.).

14 Barry S. Thornton & Philip M. Stanley, *Computers in Australia – Usage and Effects*, pp. 9–11. Dr Thornton, Dean of Mathematical and Computing Sciences at the NSW Institute of Technology, was formerly Technical Director of Honeywell Australia.

15 ibid., pp. 28–9, 46–9.

16 Australian Computer Equipment Suppliers' Association, submission to the CITCA (1979), p. 6. Note the estimate of 77 000 people employed, given earlier in the text, from the Foundation for Australian Resources.

17 Joseph Weizenbaum, *Computer Power and Human Reason: From Judgment to Calculation*, pp. 128–9.

18 Mary Wollstonecraft Shelley's *Frankenstein; or, the Modern Prometheus* (1818) is an admirable cautionary tale about the dangers of adopting a technology, not knowing what to do with it, and being unable to cope with the moral problems it raises. But Dr Frankenstein's artificial man was not a robot.

19 Norbert Wiener, 'The Brain and the Machine', in *Dimensions of the Mind.*

6 UNEMPLOYMENT, INFLATION, DEMAND AND PRODUCTIVITY

1 Dennis Altman, *Rehearsals for Change*, pp. 47–8.

2 Nripesh Podder, *The Economic Circumstances of the Poor.* This analysis was based on the Australian Survey of Consumer Finances and Expenditure 1966–68, carried out by Macquarie and Queensland Universities. The Australian Commission of Inquiry into Poverty published forty reports on the incidence of poverty in Australia.

3 See Phil Raskall, 'Who's Got What in Australia: The Distribution of Wealth' in Greg Crouch, Ted Wheelwright & Ted Wilshire (eds) *Australian and World Capitalism.*

4 'In the year to October 1974, minimum award rates for men rose by 37·2 per cent and for women by 46·1 per cent – at a time when the various price indices were increasing at around 20 per cent.' See Treasury Economic Paper no. 4, *Job Markets*, p. 33.

5 ibid., p. 94. The CES figures comprised unemployed persons registered for full-time work at CES offices. The ABS figures were derived from regular samples from a population survey of 0·6 per cent of Australian households, and estimated the numbers of people aged fifteen and over who were not employed and were actively seeking full- or part-time work.

6 ibid., p. 78.

7 ibid., p. 54.

8 Myers Report, vol. 1, p. 87.

9 Australian Bureau of Statistics, *Multiple Job Holding, Australia 1979.*

10 ABS, *Unemployment, Underemployment and Related Statistics, Australia February 1978 to February 1980*, p. 34.

11 Keith Windschuttle, *Unemployment*, pp. 144–8.

12 *Hansard*, House of Representatives, Estimates, 23 October 1979, pp. 452–4.

13 R. G. Gregory & R. Duncan, 'Participation in the Australian Labour Market: Puzzles and Conjectures', paper to Australian Agricultural Economics Society, February 1980.

14 Barry Hughes, *Exit Full Employment*, pp. 197–9.

15 In 1976, Australia manufactured more passenger cars (369 000) than Sweden (307 000).

16 Austria, Japan, Sweden and West Germany have these elements in common: they

maintain the post-war consensus style in politics and economic management, and there is heavy government involvement in the direction of investment capital (not nationalization).

17 Phillips (1914–75) was a New Zealander, an engineer turned economist who held Chairs in London and Canberra. His work was based on the relationship of prices and wages in Britain between 1860 and 1957. A weighted price index for basic commodities indicates that inflation was not a problem in England until after World War I – prices fell generally from 1813 (= 150) to 1913 (= 100), and by 1960 had risen sharply (= 480). See Phyllis Deane & W. A Cole, *British Economic Growth 1688–1959*, fig. 7.

18 J. A. Trevithick, *Inflation*, p. 56.

19 ibid., pp. 62, 64.

20 There are several increasingly expansive definitions of money supply. M1 is the total value of notes, coins and money in current account (estimated at about $14 000 million in Australia in 1980). M3 – the most often used – is M1 plus all other holdings in the banking system (about $50 000 million in 1980). M4, which is M3 plus building-society deposits, is a more useful measure.

21 Robert Bacon & Walter Eltis, *Britain's Economic Problem: Too Few Producers.* Paradoxically, Britain is in trouble with the rest of the EEC because her small agricultural labour force is too productive.

22 The Begin government in Israel, also monetarist, had an inflation rate of 140 per cent in 1980.

23 Okun (1920–80) was Professor of Economics at Yale University, advisor to the Kennedy administration, Fellow of the Brookings Institute, a prolific writer and – a rarity – an economic wit in the 'dismal science'.

24 Indecs Economics, *State of Play: Indecs Economics Special Report*, p. 18.

25 In labour/time-absorbing employment, as in the convivial sector, employment may increase in Parkinsonian fashion even where demand is stationary or falling.

26 The concept is not new – but neglected. The 'Haavelmo Effect', on which this passage is based, was first published in the article 'Multiplier Effects of a Balanced Budget' by the Norwegian economist Trygve Haavelmo, in *Econometrica* (1945).

27 'Jobs of the Future' in *Playboy*, July 1980, p. 26.

28 The inflationary tendencies of social complexity and interdependence will be greatest where they contribute to structural overload in a large unit such as a metropolis, and least where they are decentralized and dispersed.

29 *The Westudy Report,* produced by the Department of Environment, Housing and Community Development (Australian Government Publishing Service, Canberra, 1976), examined access to library and information services in Melbourne's western region. Registered library borrowers were 50 per cent higher in the east, and book borrowings per capita were 80 per cent higher.

30 The comparison understates the degree of divergence. There are 386 shops in the immediate vicinity of the Kooyong sample, and the area also has good public transport links with Melbourne's CBD. The St Albans shopping centre, relatively isolated, contains 170 shops. The Mallee sample comes from small towns in the region.

31 Ernst Engel (not to be confused with Friedrich Engels) was a Prussian statistician who stated his Law in 1857.

32 Jane Jacobs, *The Economy of Cities*, ch.3.

33 United Nations, *Yearbook of Industrial Statistics,* cited in G. T. Kurian, *The Book of World Rankings*, pp. 169–70.

34 World Bank listings, cited in Kurian, pp. 171–2. The industrial sector is defined here as mining, manufacturing, construction and utilities.

35 Overseas Development Council (Washington), cited in Kurian, pp. 286–7.

36 Peter Sheehan, *Crisis in Abundance*, pp. 103–04.

37 Peter B. Dixon & Alan A. Powell, *Structural Adaptation in an Ailing Macroeconomy*, p. 20.

38 'Aluminium Expansion and its Impact on the Australian Economy' in *Journal of Industry and Commerce* (Department of Industry and Commerce, Canberra, December 1979).

7 EDUCATION AND EMPLOYMENT

1 See R. W. Connell, *Ruling Class, Ruling Culture*, chs 7 and 8. Interviewing working-class children, he noted how readily they responded: 'I'd like a good job, but I don't have the brains.' This is reminiscent of Aldous Huxley's *Brave New World*, where the five classes are taught 'Elementary Class Consciousness' by recorded messages as they sleep: 'Alpha children wear grey. They work much harder than we do, because they're so frightfully clever . . .'
2 *Education and Poverty in Australia*, Fifth Main Report of the Australian Commission of Inquiry into Poverty, pp. 11, 227, 231.
3 See *Australian Students and their Schools* (Schools Commission, Canberra, 1979).
4 *Education and Poverty in Australia*, p. 149. Aborigines are even more disadvantaged.
5 Thomas Hardy's novel *Jude the Obscure* (1895) is the moving story of a working-man's efforts to obtain a university education.
6 *Education and Poverty in Australia*, p. 10. The Bell quotation comes from *The Coming of Post-industrial Society*, p. 453.
7 Randall Collins, *The Credential Society: An Historical Sociology of Education and Stratification* (Academic Press, New York, 1980).
8 *New York Review of Books*, 20 March 1980. Girls who have completed Year 12 generally take about six months on-the-job training to catch up to girls who have studied typing and shorthand at school for three years. IBM executives say that they can train any bright entrant in a few months, and would prefer an A-grade student in classical Greek to a B-grade student with three years of computer studies behind him.
9 These figures are for 1979, and include full- and part-time students.
10 In 1973, TAFE had a total of 205 500 students aged 17–22, of whom 15 500 were full-time. In 1977, there were 314 200 in this group, 26 200 of them full-time.
11 *Australian Students and their Schools*, pp. 61–3. (The figures in the diagram are wrong.)
12 *Hansard*, House of Representatives, 12 September 1979, vol. 115, p. 1057. Answer provided by the Minister for Education. In 1920, Australia's population was 5·3 million and there were 7900 university students (1·65 per cent of the 17–22 age group). Universities continued to double their numbers every ten years or so from 1947 until 1972 when there were 128 700 students (the equivalent of 10·75 per cent of the 17–22 age group): however, enrolment figures included many mature-age students.
13 *Hansard*, House of Representatives, 4 December 1980, p. 469. Answer provided by the Minister for Education.
14 *Education, Training and Employment*, Report of the Committee of Inquiry into Education and Training (Chairman: Sir Bruce Williams), vol. 3, s. 17, p. 7. Apart from the chairman, the committee comprised nine members drawn from education, industry and the trade-union movement. Formerly Vice-Chancellor of Sydney University, Williams resigned to become Director of the Technical Change Centre, London.
15 Free marketeers argue that state-run educational bureaucracies do not provide particularly good value for money, and have suggested that parents be issued with 'vouchers' for the equivalent of their share of taxation going to the existing state system. State schools would then compete with private schools for these voucher funds. In Australia, the voucher system (in health as well) has been advocated by Professor Richard Blandy as part of the 'libertarian scenario' set out in Wolfgang Kasper et al., *Australia at the Crossroads: Our Choices to the Year 2000* (Harcourt Brace Jovanovich, Sydney, 1980), a lively polemic against mercantilist survivals in Australian economic policy.
16 *Education, Training and Employment*, vol. 1, para. 4.21, pp. 91–2.
17 ibid., vol. 1, para. 4·31, p. 96.
18 Myers Report, vol. 1, paras 5·47. 5·51, pp. 90–1.
19 Titian, Picasso and Chagall, for example, were painting when in their nineties; Michelangelo and Henry Moore were sculpting in their eighties; and many great musicians (e.g. Segovia, Stokowski, Rubinstein, Casals, Boult) were active in their late eighties and over. Verdi wrote *Falstaff* at the age of seventy-nine.

20 The Minister for Education, Wallace Fife, reported bleakly in answer to a question on notice: 'No estimates have been made of the percentage of young people aged between 17 and 24 years who would have the intellectual capacity to benefit from tertiary education in Australia.' *Hansard*, House of Representatives, 4 December 1980, p. 469. (In other words, it appears, we do not know the degree of intellectual wastage, nor where it is found, and we do not propose to find out.)

21 Although not quoted here, Dr Shirley L. Smith's *Schooling: More or Less*, is a splendid analysis of current educational needs, with appropriate emphasis on recurrent education.

22 Quoted in Christopher Lasch, *The Culture of Narcissism*, p. 127.

23 'We live in a society which pays massive informal homage to individualism. We have structured an educational system which formally imbues our youth, year after year, with the social values of freedom, liberty and individual expression. Then they learn that they are expected to spend a lifetime on a job while explicitly submitting to authority. Even for those new labor-force entrants from the mainstream of society, it may take some time before the leisure-time substitution is made and accepted. And the rising educational level of the labor force makes the contradiction steadily more severe. But the significant point is that the contradiction itself is not only unnecessary but costly.' K. O. Alexander, 'On Work and Authority' in Leigh Estabrook (ed.), *Libraries in Post-Industrial Society* (Oryx, New York, 1977).

8 THE INFORMATION EXPLOSION AND ITS THREATS

1 Human beings are 60 per cent water – but a hydrologist does not necessarily have the greatest understanding of the human condition.

2 Langdon Winner, *Autonomous Technology*, p. 284.

3 Charles E. Lindblom, 'The Science of Muddling Through' in *Public Administration Review 19* (1959), cited in Winner, op. cit., p. 291.

4 The numbers in the House of Representatives increased to 125 in the 32nd Parliament elected in October 1980. State and municipal employees increase the numbers to nearly 1·6 million.

5 *Hansard*, House of Representatives, 19 February 1980, pp. 83–88. Answer by Finance Minister, Eric Robinson, to a question on notice. The full list of 148 Acts was given and the answer took five months to prepare. Not one newspaper picked up its significance.

6 *Hansard*, House of Representatives, 7 June 1979, vol. 114, p. 3158. Answer by the Prime Minister, Malcolm Fraser, to a question on notice.

7 *National Times*, 28 October 1978.

8 See Bertram M. Gross, *Friendly Fascism: The New Face of Power in America*. The book received a very adverse review from Jason Epstein in the *New York Review of Books*, 23 October 1980.

9 Mandata is a computerized system for processing manpower data in the Commonwealth public service. Its long, complex and confusing history is described in the Public Accounts Committee Report 175, 1979. For details of the Pharmaceutical Benefits Scheme, see Public Accounts Committee Report 182, 1980.

10 Fritz Machlup, *The Production and Distribution of Knowledge in the United States*.

11 D. M. Lamberton (ed.), *Economics of Information and Knowledge*.

12 Daniel Bell, 'Teletext & Technology' in *Encounter*, June 1977.

13 *Information in 1985: A Forecasting Study of Information Needs and Resources* (OECD, Paris, 1973).

14 Quantity and quality of information are by no means synonymous. The vast bulk of material processed and stored may be of no use to anyone. Computer programmers have coined the useful acronym GIGO – 'garbage in: garbage out'. Data are raw material, and information is organized, selected, data in a usable form (thus D → I). However, when a massive volume of information is presented in such a form that it becomes useless, its value is reduced to that of data, i.e. I → D. This is called 'information overload'.

15 George A. Miller, *The Psychology of Communication: Seven Essays*.

16 This is not so. Information can be, and often is, suppressed, lost, erased or shredded.
17 Yoneji Masudi, 'Future Perspectives for Information Utility'.
18 Martin & Norman, op. cit., pp. 29, 31.
19 Some existing examples are Euronet (European area), Tymnet (US), Orbit (international), Diolog (international) and Ausinet (Australasia).
20 *The Computerisation of Society*, pp. 49–50.
21 Masuda, op. cit.
22 *The Computerisation of Society*, pp. 27–8, 52–3. (I have slightly amended the translation for greater clarity.)
23 *Hansard*, House of Representatives, 18 October 1979, pp. 2224–42. Telecom was vigorously opposed to any threat to its monopoly status as a communications common carrier. The question of who would control the satellite was side-stepped by the Fraser government.
24 During World War II, the Germans exterminated a greater proportion of Dutch Jews than French Jews because personal data from censuses and other official records was far more accurate and complete in the Netherlands than in France.
25 Martin & Norman, op. cit., p. 363.
26 See s. 51 (v).

9 WORK IN AN AGE OF AUTOMATA

1 Genesis iii, 19. The curse of Adam is referred to nine times in the Old Testament and Apocrypha. In the New Testament (II Thessalonians iii, 12), St Paul's command and exhortation to 'disorderly . . . busybodies . . . that with quietness they work, and eat their own bread' is regarded as the only reference to 'the curse'.
2 A. W. Zimmern, *The Greek Commonwealth.*
3 Marx paraphrased this passage (inaccurately) in *Capital*, vol. 1, ch. 15.
4 The story of Jesus, Martha and Mary (Luke x, 38–42) lends no support to the work ethic. However, the parable of the talents (Matthew xxv, 14–39) is in the authentic spirit of capitalism.
5 See II Thessalonians iii, 7–10. In *The Jerusalem Bible* (London, 1966) the last verse is rendered 'We gave you a rule when we were with you: not to let anyone have food if he refused to do any work', and a footnote comments: 'This may have been laid down by Jesus, but it may have been a proverb: it has been called the golden rule for Christian work.'
6 P. D. Anthony, *The Ideology of Work*, p. 22.
7 Jérôme Carcopino, *Daily Life in Ancient Rome.*
8 Sol Encel, 'The Future of Work, Education and Leisure', paper delivered to the Australian and New Zealand Association for the Advancement of Science, May 1979.
9 Emmanuel Le Roy Ladurie, *Montaillou,* p. 339.
10 Quoted in Anthony, op. cit., p. 40.
11 Octavio Paz, 'Mexico and the United States', in *New Yorker,* 17 September 1979, pp. 144–5.
12 The Swedish company Volvo is an exception: its cars are built on a craftwork basis by teams of highly skilled workers – but they do not make the components. Rolls Royce is another example, but with smaller production, higher craft element and greater cost.
13 Braverman, op. cit., pp. 272–3, 276.
14 ibid. p. 53.
15 *Economic and Philosophical Manuscripts,* written in Paris in 1844. They were not published until 1932, and are considered as the work of a romantic humanist rather than a scientific socialist. The original concept of 'alienation' comes from Rousseau.
16 See *Grundrisse* (tr. Martin Nicolaus), p. 611.
17 *Capital* (1974), p. 920.
18 ibid., pp. 386–7.
19 David McLellan (tr.), *Marx's Grundrisse*, pp. 154–6, 162–3. The McLellan translation is less confusing than the Penguin edition at this point. The quotation at the end is from Goethe.

20 ibid. Ch. 15 of *Capital* (vol. 1) is an encyclopedic account of the past impact of machinery and large-scale industry on employment, but it does not touch on automation or predict the future.

21 Marx has also been attacked as 'the apotheosis of bourgeois materialism' for his emphasis on man as an economic animal. He sometimes propounded astonishingly rigid views, e.g. in 'Critique of the Gotha Programme' (1875) where he attacked the German Social Democrats for including the prohibition of child labour in their policies. (He *may* have intended by this, however, that the existence of rules to protect child workers could be the thin edge of the wedge to extend safety and wage guarantees to other workers in the future.)

22 Marx himself was not so dogmatic – but as he told Paul Lafargue about the French communists: 'What is certain is that I am no Marxist' (quoted by Engels in a letter to Eduard Bernstein). Lafargue, Marx's son-in-law, rejected the work ethic, asserting that workers should have *le droit à paresse* (the right to idleness).

23 The traditional work ethic dies hard. 'Beyond Unemployment: A Statement on Human Labour' prepared by the Catholic Commission for Justice and Peace for the Catholic bishops of Australia, was published in 1979. It urged the abolition of the work test imposed on those seeking unemployment benefits, raising those payments to the poverty line, and re-defining work to include unpaid community service. The report was roundly denounced as an attempt to undermine the work ethic and most bishops repudiated it. A 1980 New Zealand survey on the quality of life, carried out in Dunedin, found that 20 per cent of those questioned thought the unemployed were 'not entitled to any quality of life' and 60 per cent thought they were entitled only to 'a very low standard of living'.

24 Bertrand Russell, op. cit., pp. 12, 14.

25 Young and Willmott, op. cit. Loss of a 'work faculty' may be equated to losing an arm: it is of little consolation to the amputee that he is adequately provided for with a pension or lump-sum payment.

26 Ironically, Japan is a gerontocracy at the management level and in politics, despite the youth of its labour force. This is in sharp contrast to Australia, where there is an advanced retirement age and a youthful parliament. In the US, advocates of 'grey power' are opposed to mandatory retirement ages – as, presumably, is President Reagan.

27 Martin & Norman, op. cit., pp. 455, 491.

28 Henderson Report, vol. 1.

29 The Nixon administration attempted to introduce a guaranteed income scheme in 1969. This was defeated in the Senate by a combination of conservatives (who thought the plan was contrary to the work ethic) and radicals (who wanted a more generous scheme). The controversy is described brilliantly in *The Politics of a Guaranteed Income: The Nixon Administration and the Family Assistance Plan* by Daniel Patrick Moynihan. See also the long review article, 'Who Crushed Nixon's Revolution?' by Tom Wicker in the *New York Review of Books*, 22 March 1973.

30 The GAMMA Report, vol. 4, pp. 14–17.

31 Edward de Bono calls this 'non-market labour' (NML) in *Future Positive*, p. 161.

32 I asked the Minister for Productivity in a question on notice: Is he able to state whether the introduction of the 40-hour week in Australia in 1947–48 had any adverse effects on productivity; if so, can he state what they were? Further, can he state if the introduction of the three-day week by the Heath government in 1973–74 had any adverse effects on production? The answer to each question was 'No'. *Hansard*, House of Representatives, 19 August 1980, pp. 452–3.

33 Archibald A. Evans, *Hours of Work in Industrialised Countries*, Appendix II, table I.

34 On the other hand, small branches in some banks have been able to enlarge ranges of service with minimal staff and computerized technology.

10 TECHNOLOGICAL DETERMINISM: 'BUT WE HAVE NO CHOICE'

1 Violent protest against technology was an important element in Japan's Satsuma rebellion (1877) against the Westernizing tendencies of the Meiji Restoration: its leader Saigo Takamori is vividly described in Ivan Morris' *The Nobility of Failure*. The

Chinese Cultural Revolution and the rule of the 'Gang of Four', and the Islamic Revolution in Iran, were similarly motivated.

2 Kimon Valaskasis, Peter S. Sindell, J. Graham Smith & Iris Fitzpatrick-Martin, *The Conserver Society: A Workable Alternative for the Future*, pp. 16–22.

3 Lewis Mumford, *The Myth of the Machine*, pp. 5–6, 9.

4 Robert K. Merton, in Ellul, op. cit., p. viii.

5 Nigel Calder has pointed out in his *Technopolis: Social Control of the Uses of Science*, that a project to abolish darkness by putting a one-kilometre-wide mirror into orbit around the earth was rejected by the US National Academy of Sciences.

6 Ellul, op. cit., p. xxxiii.

7 Langdon Winner, *Autonomous Technology*, pp. 229, 234, 237, 323–4.

8 Barry Commoner, *The Poverty of Power*, pp. 177–9.

9 By offering Comalco the cheapest electricity in the world (0.6 cents per kwh) and passing the cost on to domestic consumers and labour-intensive industries, the HEC makes it difficult for small-scale industries to survive.

10 Albert Speer, *Inside the Third Reich*, pp. 211, 212.

11 From the transcript 'In the Matter of J. Robert Oppenheimer', in Winner, op. cit., p. 73.

12 Nicolas Freeling's novel *Gadget* (1979) illustrates the technical fascination of atomic weaponry very plausibly. A middle-ranking engineer with experience in atomic plants is kidnapped by terrorists who try to force him to design a low-yield atomic bomb. At first he refuses, but the excitement of the challenge soon overcomes his moral revulsion. Colonel Nicholson, in the film *Bridge over the River Kwai*, had the same response.

13 Cited in Mumford, *The Pentagon of Power*, pp. 280–1. The US Supreme Court's judgement in the *Chakrabarty* case (1980) – that biotechnological developments could be patented – raises the question of heavy investment in, and pressure to adopt, genetic engineering.

14 Mumford, op. cit., pp. 185–6.

15 ibid., p. 173.

16 Weizenbaum, op. cit., pp. 255–6, 256–7.

17 Conversation with author, May 1979. Dr Krishan Kumar, of the University at Canterbury, author of *Prophecy and Progress* (Penguin, 1978), contends in an essay 'Thoughts on the Present Discontents in Britain: A Review and a Proposal' (*Theory and Society*, vol. 9/3, Amsterdam, 1980), and not ironically, that Britain is – once again – an innovator in a new development of the Industrial Revolution, with greater emphasis on quality of life, leisure and a reduced emphasis on materialism. Professor John Kenneth Galbraith supports this position.

18 In the parking area outside Parliament House in Canberra, there are normally over thirty Volvos or Saabs. Holdens are rarely found in the Riksdag's car park in Stockholm.

19 Myers Report, vol. 1, p. 77. The words 'multi-national corporation' do not appear in the report.

20 The *Australian Financial Review* has printed only one side of the 'technology debate' and completely ignored, for example, the Nora Report. In September 1980, sixty of its 1180 pages were devoted to promoting computers.

21 Kahn and Pepper, op. cit., p. 186. Kahn was founder of the Hudson Institute, a policy-research 'think tank' in New York.

22 Connell, op. cit., ch.4. CRA (Conzinc Riotinto of Australia) was a successor company to the Zinc Corporation, founded in 1905. CRA owns a 45 per cent share of Comalco.

23 *Industrial Research and Development*, Report of the Senate Standing Committee on Science and the Environment (Australian Government Publishing Service, Canberra, 1979).

24 Mike Cooley, *Architect or Bee?*, pp. 63–82.

25 See *Energy Options and Employment* (Centre for Alternative Industrial & Technological Systems, North-East London Polytechnic, 1979), pp. 33–5.

26 See *The Efficient Use of Energy* (The American Physical Society 1975).

27 New Zealand Planning Act 1977, no. 76.

28 *Washington Post*, 23 May 1977, quoted in Hazel Henderson, *Creating Alternative Futures*, p. 395.
29 Dr John Morris, 'Changing Technology in Hospitals', Proceedings of the Australian Hospital Association Annual Congress, Canberra, 1979.
30 Australian Royal Commission of Inquiry into Drugs (Commissioner: Mr Justice Sir Edward Williams; Australian Government Publishing Service, Canberra, 1980).
31 In Australia, Commonwealth expenditure on solar-energy research amounted to less than $3 million for 1980–81.
32 Ron Johnston & Philip Gummett (eds), *Directing Technology*, pp. 196–7.
33 Barry W. Smith, 'The Social and Economic Consequences of Technological Change', paper prepared for the CITCA, 1979.
34 The Economic Research Group of the Commonwealth Parliamentary Library was unable to come up with a firm figure, but concluded that it was between 5 and 10 per cent.
35 See Myers Report, vol. 1, para. 4·77.
36 Paul Harwitz, in *The Wall Street Journal*, quoted in *Omni*, February 1980.

11 WHAT IS TO BE DONE?

1 One example is the equine culture: there are now more horses kept for breeding, showing and schooling than at any time in Australia's history.
2 In the *New York Review of Books*, 5 February 1981. By 1980, McDonald's fast-food chain had more employees than US Steel. There are more hairdressers in the USA than steel-workers.
3 Alternatively, 'occupation' – which relates to time use rather than income – could be used instead of 'work'.
4 The Commonwealth Departments of Education, Productivity, and Employment and Youth Affairs lack adequate statistical bases for developing appropriate social policies.
5 Legislation will not automatically produce full employment – but it indicates that the community accepts full responsibility for achieving the goal.
6 See *Freedom of Information*, Report of the Senate Constitutional and Legal Committee (Australian Government Publishing Service, Canberra, 1979).
7 Report of the Royal Commission on Australian Government Administration (Chairman: Dr H. C. Coombs; Australian Government Publishing Service, Canberra, 1976).
8 This credulity sometimes extends to scientists outside their own discipline – witness the support given by some physicists to Uri Geller's spoon-bending, in contrast to professional magicians and conjurors who were deeply sceptical.
9 There is a passionate attack on reductionism in Robert Pirsig's *Zen and the Art of Motorcycle Maintenance* (Corgi, London, 1974).

Bibliography

Date of original publication of some works is indicated by square brackets, [ ]: the most recent and/or accessible edition is listed.

Allaby, Michael, *Inventing Tomorrow*, Abacus, London, 1977.

Altman, Dennis, *Rehearsals for Change*, Fontana, Melbourne, 1980.

Anthony, P. D., *The Ideology of Work*, Tavistock, London, 1977.

Arendt, Hannah, *The Human Condition*, University of Chicago Press, Chicago, 1958.

Aron, Raymond, *The Industrial Society*, Simon & Schuster, New York, 1967.

Asimov, Isaac, *Biographical Encyclopedia of Science and Technology*, Pan, London, 1974.

Australian Academy of Science, *The Impact of Microprocessors on Industry, Education and Society*, (ed. James D. Morrison), Canberra, 1980.

Australian and New Zealand Association for the Advancement of Science, *Automation and Employment*, Law Book Co., Sydney, 1979.

Australian Council of Trade Unions – Council of Australian Government Employees' Organisations (ACTU–CAGEO), *Joint Peak Union Council Submission to CITCA* (ed. Sandra Prerost), Melbourne, 1979.

Australian Schools Commission, *Australian Students and their Schools*, Canberra, 1979.

Australian Senate Constitutional and Legal Committee, *Freedom of Information*, Australian Government Publishing Service, Canberra, 1979.

Australian Senate Standing Committee on Science and the Environment, *Industrial Research and Development*, Australian Government Publishing Service, Canberra, 1979.

Australian Treasury, *Job Markets*, Treasury Economic Paper no. 4, Australian Government Publishing Service, Canberra, 1979.

——, *Technology, Growth and Jobs*, Treasury Economic Paper no. 7, Australian Government Publishing Service, Canberra, 1979.

Babbage, Charles, *On the Economy of Machinery and Manufactures*, Charles Knight, London, 1832.

Bacon, Robert & Eltis, Walter, *Britain's Economic Problem: Too Few Producers*, Macmillan, London, 1976.

Baran, Paul A., *The Political Economy of Growth*, Penguin, London, 1973.

Baran, Paul A. & Sweezy, Paul M., *Monopoly Capital*, Penguin, London, 1977.

Barnet, Richard J. & Müller, Ronald E., *Global Reach: The Power of the Multinational Corporations*, Simon & Schuster, New York, 1974.

Becker, Gary S., 'A Theory of the Allocation of Time', in *Economic Journal*, September 1965.

Bell, Daniel, *The Coming of Post-Industrial Society: A Venture in Social Forecasting*, Heinemann, London, 1974.

——, 'Teletext and Technology', in *Encounter*, June 1977.

Bernal, J. D., *Science in History*, vols I–IV, Penguin, London, 1977.

Blainey, Geoffrey, *The Causes of War*, Macmillan, London, 1973.

Blandy, Richard et al., *Australia at the Crossroads: Our Choices to the Year 2000*, Harcourt Brace Jovanovich, Sydney, 1980.

Braverman, Harry, *Labor and Monopoly Capital: The Degradation of Work in the Twentieth Century*, Monthly Review Press, New York, 1974.

Burnham, Jack, *Beyond Modern Sculpture*, Allen Lane, London, 1968.

Burns, Scott, *Home, Inc.*, Doubleday, New York, 1975.

Cairns, Jim, *Growth to Freedom*, Down to Earth, Canberra, 1979.

Calder, Nigel, *Technopolis: Social Control of the Uses of Science*, Simon & Schuster, New York, 1971.

Calvocoressi, Peter, *The British Experience 1945–75*, Penguin, London, 1979.

Čapek, Karel, *R.U.R. and the Insect Play* [1921], Oxford University Press, Oxford, 1973.

Carcopino, Jérôme, *Daily Life in Ancient Rome*, Penguin, London, 1956.

Central Policy Review Staff, *Social and Economic Implications of Microelectronics*, London, 1978.

Chandor, Anthony, et al., *A Dictionary of Computers*, Penguin, London, 1977.

Cipolla, Carlo M. (ed.), *The Fontana Economic History of Europe*, vols III and IV, Fontana/Collins, London, 1973.

——, *The Economic History of World Population*, Penguin, London, 1962.

——, *Literacy and Development*, Penguin, London, 1969.

Clark, Colin, *The Conditions of Economic Progress*, Macmillan, London, 1940.

Clarke, Arthur C., *Profiles of the Future*, Pan, London, 1964.

Clarke, Robin, *Notes for the Future*, Thames & Hudson, London, 1975.

Club of Rome, *The Limits to Growth* (ed. D. H. Meadows), New American Library, New York, 1974.

——, *Mankind at the Turning Point* (eds. Mihajlo Mesarovic & Eduard Pestel), Hutchinson, London, 1975.

Cohen, Barry, *Green Paper on Sport and Recreation*, Australian Labor Party, Canberra, 1980.

Commoner, Barry, *The Poverty of Power*, Bantam, New York, 1977.

Cooley, Mike, *Architect or Bee?*, Langley, Slough, 1980.

Coombs, Herbert Cole, et al., *Report of the Royal Commission on Australian Government Administration*, vols I–III, Australian Government Publishing Service, Canberra, 1976.

Connell, R. W., *Ruling Class, Ruling Culture*, Cambridge University Press, Cambridge, 1977.

Connell, W. F., *A History of Education in the Twentieth Century World*, Prentice-Hall, Sydney, 1980.

Grough, Greg, et al., *Australia and World Capitalism*, Penguin, Melbourne, 1980.

Davison, Graeme, *The Rise and Fall of Marvellous Melbourne*, Melbourne University Press, Melbourne, 1978.

Deane, Phyllis & Cole, W. A., *British Economic Growth 1688–1959*, Cambridge University Press, Cambridge, 1967.

de Bono, Edward, *Future Positive*, Penguin, London, 1980.

Dertouzos, Michael & Moses, Joel (eds), *The Computer Age*, MIT Press, Cambridge, Mass., 1980.

Dickson, David, *Alternative Technology*, Fontana/Collins, London, 1974.

Dixon, Peter B. & Powell, Alan A., *Structural Adaptation in an Ailing Macro-economy*, Melbourne University Press, Melbourne, 1979.

Drucker, Peter, *The Age of Discontinuity*, Heinemann, London, 1969.

Dubos, René, *A God Within*, Abacus, London, 1976.

Dyson, Freeman, *Disturbing the Universe*, Harper & Row, New York, 1979.

Ellul, Jacques, *The Technological Society* (tr. John Wilkinson), Knopf, New York, 1964.

——, *The Technological System*, Continuum, New York, 1980.

Estabrook, Leigh (ed.), *Libraries in Post-Industrial Society*, Oryx, New York, 1977.

Evans, Archibald A., *Hours of Work in Industrialised Countries*, International Labour Office, Geneva, 1975.

Evans, Christopher, *The Mighty Micro*, Hutchinson/Gollancz, London, 1979.

Fisher, A. G. B., *The Clash of Progress and Security*, Macmillan, London, 1935.

Fitzgerald, Ronald T., *Education and Poverty in Australia*, Fifth Main Report of the Commission of Inquiry into Poverty, Australian Government Publishing Service, Canberra, 1976.

Forester, Tom (ed.), *The Microelectronics Revolution*, Basil Blackwell, Oxford, 1980.

Forster, E. M., 'The Machine Stops' (1908), in *Collected Short Stories*, Penguin, London, 1954.

Fourastié, Jean, *Le grand espoir du XXe Siècle*, Presses Universitaires de France, Paris, 1949.

Freeling, Nicolas, *Gadget*, Penguin, London, 1979.

Freud, Sigmund, *Civilization and its Discontents*, Hogarth, London, 1953.

Friedman, Milton & Rose, *Free to Choose*, Penguin, London, 1980.

Gabor, Dennis, *The Mature Society*, Praeger, New York, 1972.

——, *Inventing the Future*, Penguin, London, 1964.

Galbraith, John Kenneth, *The Affluent Society*, Penguin, London, 1962.

——, *The New Industrial State*, Penguin, London, 1969.

——, *Almost Everyone's Guide to Economics* (with Nicole Salinger), Houghton Mifflin, Boston, 1978.

——, *Annals of an Abiding Liberal*, André Deutsch, London, 1980.

Galenson, W. & Zellner, A., 'International Comparison of Unemployment Rates', in *The Measurement and Behaviour of Unemployment*, Universities–National Bureau Committee for Economic Research, Princeton, 1957.

GAMMA Report, *The Selective Conserver Society*, vols I–IV, McGill and

Montreal Universities, 1977.
Gershuny, Jonathan, *After Industrial Society?*, Macmillan, London, 1978.
Gilder, George, *Wealth and Poverty*, Basic Books, New York, 1981.
Gilmore, Peter & Lansbury, Russell, *Ticket to Nowhere*, Penguin, Melbourne, 1978.
Gimpel, Jean, *The Medieval Machine*, Futura, London, 1979.
Ginzberg, Eli & Vojta, George J., 'The Service Sector of the US Economy', in *Scientific American*, March 1981.
Gough, Ian, *The Political Economy of the Welfare State*, Macmillan, London, 1979.
Gregory, R. G. & Duncan, R., 'Participation in the Australian Labour Market: Puzzles and Conjectures', paper to the Australian Agricultural Economics Society, February 1980.
Gregory, R. H., 'Some Implications of the Growth of the Mineral Sector', in *The Australian Journal of Agricultural Economics*, vol. 20, no. 2, August 1976.
Gross, Bertram M., 'Planning in an Era of Social Revolution', in *Public Administration Review*, May–June 1971.
——, *Friendly Fascism*, Evans, New York, 1980.
Hainsworth, Frank, *Economics*, vols I and II, Methuen, Sydney, 1978.
Hancock, K. J., et al., *A National Superannuation Scheme for Australia*, Report of the National Superannuation Committee of Inquiry, Australian Government Publishing Service, Canberra, 1976.
Hardy, Thomas, *Jude the Obscure* [1895], Macmillan, London, 1974.
Hartwell, R. M., 'The Service Revolution: The Growth of Services in Modern Economy 1700–1914', in C. M. Cipolla (ed.), *Fontana Economic History of Europe*, vol. 3, Collins, London, 1973.
Heilbroner, Robert L., *An Inquiry into the Human Prospect*, Norton, New York, 1974.
——, *Business Civilisation in Decline*, Penguin, London, 1976.
Henderson, Hazel, *Creating Alternative Futures*, Berkley, New York, 1975.
Henderson, Ronald, et al., *Poverty in Australia*, First Main Report of the Commission of Inquiry into Poverty, Australian Government Publishing Service, Canberra, 1975.
Higgins, Ronald, *The Seventh Enemy*, Pan, London, 1980.
Hirsch, Fred, *Social Limits to Growth*, Routledge & Kegan Paul, London, 1977.
Hobsbawm, Eric John, *Industry and Empire*, Penguin, London, 1968.
——, *The Age of Capital 1848–1875*, Abacus, London, 1977.
——, *The Age of Revolution 1789–1848*, Cardinal, London, 1973.
Hofstadter, Douglas J., *Gödel, Escher, Bach: An Eternal Golden Braid*, Penguin, London, 1980.
Holland, Stuart, *The Socialist Challenge*, Quartet, London, 1975.
Hughes, Barry, *Exit Full Employment*, Angus & Robertson, Sydney, 1980.
Huxley, Aldous, *Brave New World* [1932], Penguin, London, 1955.
Illich, Ivan D., *De-Schooling Society*, Penguin, London, 1971.
——, *Celebration of Awareness*, Penguin, London, 1976.
——, *Tools for Conviviality*, Fontana, London, 1975.

——, *Energy and Equity*, Calder & Boyars, London, 1973.

——, *The Right to Useful Unemployment*, Calder & Boyars, London, 1979.

——, *Medical Nemesis*, Calder & Boyars, London, 1974.

Indecs Economics, *State of Play: Indecs Economics Special Report*, Allen & Unwin, Sydney, 1980.

Jackson, R. G., et al., *Policies for Manufacturing Industry*, Report of Committee to Advise on Policies for Manufacturing Industry, 3 vols, Australian Government Publishing Service, Canberra, 1976.

Jacobs, Jane, *The Economy of Cities*, Penguin, London, 1972.

Jenkins, Clive & Sherman, Barrie, *The Collapse of Work*, Eyre Methuen, London, 1979.

Johnston, Ron & Gummett, Philip (eds), *Directing Technology*, Croom Helm, London, 1979.

Jones, Barry O., 'Implications of a Post Industrial or Post Service Revolution on the Nature of Work', submission to CITCA, 1979.

——, 'The Social Impact of Microcomputers', in Australian Academy of Science, *The Impact of Microprocessors on Industry, Education and Society* (ed. James D. Morrison), Canberra, 1980.

Kahn, Herman & Pepper, Thomas, *Will She Be Right? The Future of Australia*, University of Queensland Press, Brisbane, 1980.

Kalecki, Michal, *Studies in the Theory of Business Cycles 1933–39*, Basil Blackwell, Oxford, 1966.

Kemp, David A., *Society and Electoral Behaviour in Australia*, University of Queensland Press, Brisbane, 1978.

Keynes, John Maynard, *Essays in Persuasion*, Macmillan, London, 1931.

——, *The General Theory of Employment, Interest and Money*, Macmillan, London, 1936.

Kondratiev, Nikolai D., 'The Long Waves in Economic Life', in *Readings in Business Cycle Theory*, American Economic Association, 1950.

Kozinski, Jerzy, *Being There*, Corgi, London, 1980.

Kuhns, William, *The Post-Industrial Prophets*, Harper, New York, 1973.

Kumar, Krishan, *Prophecy and Progress*, Penguin, London, 1978.

——, 'Thoughts on the Present Discontents in Britain', in *Theory and Society*, vol. 9, no. 4, July 1980.

Kurian, George T., *The Book of World Rankings*, Macmillan, London, 1979.

Kuznets, Simon, *Modern Economic Growth: Rate, Structure and Spread*, Yale University Press, New Haven, 1966.

Lamberton, D. M. (ed.), *Economics of Information and Knowledge*, Penguin, London, 1971.

Landes, David, S., *The Unbound Prometheus*, Cambridge University Press, Cambridge, 1969.

Lansbury, Russell (ed.), *Democracy in the Work Place*, Longman Cheshire, Melbourne, 1980.

——, & Prideaux, Geoffrey J., *Improving the Quality of Work Life*, Longman Cheshire, Melbourne, 1978.

Large, Peter, *The Micro Revolution*, Fontana/Collins, London, 1980.

Lasch, Christopher, *The Culture of Narcissism*, Abacus, London, 1980.

Leakey, Richard E. & Lewin, Roger, *Origins*, Macdonald & Jane's, London 1977.

Lenin, V. I., *Selected Works*, Progress Publishers, Moscow, 1971.

Leontieff, Wassily, 'Issues of the Coming Years', in *Economic Impact*, no. 24, 1978–9.

Le Roy Ladurie, Emmanuel, *Montaillou*, Penguin, London, 1980.

Lewis, Sir W. Arthur, *The Evolution of the International Economic Order*, Princeton University Press, New Jersey, 1978.

Linge, G. J. R., *Industrial Awakening: A Geography of Australian Manufacturing 1788–1890*, Australian National University Press, Canberra, 1979.

McEvedy, Colin & Jones, Richard, *Atlas of World Population History*, Penguin, London, 1978.

McLellan, David (ed.), *Marx's Grundrisse*, Paladin, London, 1971.

——, *Karl Marx: His Life and Thought*, Paladin, London, 1976.

——, *Karl Marx: Selected Writings*, Oxford University Press, Oxford, 1977.

McLuhan, Marshall, *The Medium is the Massage*, Routledge & Kegan Paul, London, 1967.

Machlup, Fritz, *The Production and Distribution of Knowledge in the United States*, Princeton University Press, New Jersey, 1962.

Maddock, Sir Ieuan, 'The Future of Work', in *New Scientist*, 23 November 1978.

Mapperson, Ieuan, *The Implications of Techological Change*, Committee for Economic Development of Australia, Melbourne, 1979.

Marcuse, Herbert, *One-Dimensional Man*, Abacus, London, 1972.

Martin, James & Norman, Adrian, *The Computerized Society*, Penguin, London, 1973.

Marx, Karl, *The Communist Manifesto* [1848], Penguin, London, 1965.

——, *Capital*, vol. I [1867], (tr. Ben Fowkes), Penguin, London, 1976.

——, *Capital*, vol. II [1885], (tr. David Fernbach), Penguin, London, 1976.

——, *Capital*, vol. III [1894], Progress Publishers, Moscow, 1974.

——, *Grundrisse* [1939], (tr. Martin Nicolaus), Penguin, 1973.

Masuda, Yoneji, *Social Impact of Computerization: An Application of the Pattern Model for Industrial Society*, Kodansha, Tokyo, 1972.

——, 'Future Perspectives for Information Utility', in International Council for Computer Communications, *Evolution in Computer Communications*, North Holland, Amsterdam, 1978.

Mayer, Henry & Nelson, Helen (eds), *Australian Politics: A Fifth Reader*, Longman Cheshire, Melbourne, 1980.

Mayhew, Henry, *London Labour and the London Poor* [1851], Dover, London, 1969.

Miller, George A., *The Psychology of Communication: Seven Essays*, Basic Books, New York, 1967.

Mishan, E. J., *The Costs of Economic Growth*, Penguin, London, 1969.

——, *The Economic Growth Debate*, Allen & Unwin, London, 1977.

Mitchell, B. R. & Deane, Phyllis, *Abstract of British Historical Statistics*, Cambridge University Press, Cambridge, 1962.

Morris, Ivan, *The Nobility of Failure*, Penguin, London, 1980.

Moynihan, Daniel Patrick, *The Politics of a Guaranteed Income*, Random House, New York, 1973.

Mumford, Lewis, *The Myth of the Machine*, Secker & Warburg, London, 1967.

——, *The Pentagon of Power*, Secker & Warburg, London, 1970.

Myers, Sir Rupert, et al., *Technological Change in Australia*, Report of the Committee of Inquiry into Technological Change in Australia (CITCA), 4 vols, Australian Government Publishing Service, Canberra, 1980.

National Commission on Technology, Automation and Economic Progress, *Technology and the American Economy*, Washington DC, 1966.

Nesbit, Edith, *The Magic City* [1910], Macmillan, London, 1980.

Neumann, John von & Morgenstern, Oscar, *Theory of Games and Economic Behaviour*, Princeton University Press, New Jersey, 1980.

Nora, Simon & Minc, Alain, *L'informatisation de la société*, 5 volumes, La Documentation Française, Paris, 1978. Vol. I has been translated as *The Computerisation of Society*, MIT Press, Cambridge, Mass., 1980.

O'Connor, James, *The Fiscal Crisis of the State*, St Martin's, New York, 1973.

OECD, *Technical Change and Economic Policy*, Paris, 1980.

——, *Information in 1985: A Forecasting Study of Information Needs and Resources*, Paris, 1973.

——, *Australia – Economic Survey*, Paris, 1980.

——, *Conference on Computer/Telecommunications Policy*, Paris, 1975.

Orwell, George, *1984* [1949], Penguin, London, 1954.

Parker, Stanley, *The Future of Work and Leisure*, Paladin, London, 1972.

Parkinson, C. Northcote, *Parkinson's Law* [1957], Penguin, London, 1965.

Pawley, Martin, *The Private Future*, Random House, New York, 1974.

Paz, Octavio, 'Mexico and the United States', in *New Yorker*, 17 September 1979.

Pedler, Kit, *The Quest for Gaia*, Souvenir Press, London, 1979.

Pirsig, Robert, *Zen and the Art of Motorcycle Maintenance*, Corgi, London, 1974.

Playford, John & Kirsner, Douglas (eds), *Australian Capitalism*, Penguin, Melbourne, 1972.

Podder, Nripesh, *The Economic Circumstances of the Poor*, Australian Commission of Inquiry into Poverty, Australian Government Publishing Service, Canberra, 1978.

Porat, Marc Uri, *The Information Economy*, 9 vols, US Department of Commerce, 1977.

Pyke, Magnus, *There and Back*, Pan, London, 1980.

Reichardt, Jasia, *Robots: Fact, Fiction and Prediction*, Penguin, London, 1978.

Richta, Radovan, *Civilization at the Crossroads*, Czechoslovak Academy of Sciences, Prague, 1967.

Robinson, Joan, *The Economics of Imperfect Competition* (2nd ed.), Macmillan, 1969.

Rosenberg, Nathan, 'Problems in the Economists's Conceptualisation of Technological Innovation', in *History of Political Economy*, vol. 7, no. 4, 1975.

Rosenfeld, Albert, *Prolongevity*, Knopf, New York, 1976.

Rostow, Walt W., *Economics of Take Off into Sustained Growth*, Macmillan, London, 1963.

Rothschild, Emma, 'Reagan and the Real America', in *New York Review of Books*, 5 February 1981.

Rubinstein, David (ed.), *Education and Equality*, Penguin, London, 1980.

Russell, Bertrand, *In Praise of Idleness and Other Essays*, Allen & Unwin, London, 1976.

Sauvy, Alfred, *General Theory of Population*, Methuen, London, 1974.

Schon, Donald A., *Beyond the Stable State*, Penguin, London, 1973.

Schumacher, E. F., *Small is Beautiful: A Study of Economics as if People Mattered*, Abacus, London, 1974.

——, *Good Work*, Abacus, London, 1980.

——, *A Guide for the Perplexed*, Abacus, London, 1979.

Schumpeter, Joseph A., *Business Cycles: A Theoretical, Historical and Statistical Analysis of the Capitalist Process*, vols I–II, McGraw-Hill, New York, 1939.

Scott, Donald, *The Psychology of Work*, Duckworth, London, 1980.

Sheehan, Peter, *Crisis in Abundance*, Penguin, Melbourne, 1980.

Simon, Herbert A., *The Sciences of the Artificial*, MIT Press, Cambridge, Mass., 1970.

Simon, Herbert A. & Newell, Alan, *Human Problem Solving*, Prentice-Hall, New Jersey, 1972.

Sivard, Ruth Leger, *World Military and Social Priorities*, World Priorities Inc., New York, 1979.

Sleigh, Jonathan, et al., *The Manpower Implications of Micro-Electronic Technology*, Her Majesty's Stationery Office, London, 1979.

Smith, Adam, *An Inquiry into the Nature and Causes of the Wealth of Nations*, 4 vols [1776]. Most accessible in the one-vol. abridgement, *The Wealth of Nations* (ed. Andrew Skinner), Penguin, London, 1976.

Smith, Shirley L., *Schooling: More or Less*, Jacaranda, Brisbane, 1980.

Speer, Albert, *Inside the Third Reich*, Weidenfeld & Nicholson, London, 1970.

Stretton, Hugh, *Capitalism, Socialism and the Environment*, Cambridge University Press, Cambridge, 1976.

Stubbs, Peter, *Technology and Australia's Future: Industry and International Competitiveness*, Australian Industries Development Association, Canberra, 1980.

Tawney, R. H., *Religion and the Rise of Capitalism*, Penguin, London, 1948.

Taylor, Frederick Winslow, *The Principles of Scientific Management* [1911], Harper & Row, London, 1964.

Theophanous, Andrew C., *Australian Democracy in Crisis*, Oxford University Press, Melbourne, 1980.

Thomis, Malcolm I., *The Luddites: Machine Breaking in Regency England*, David & Charles, Newton Abbot, 1975.

Thompson, E. P., *The Making of the English Working Class*, Penguin, London, 1963.

Thornton, B. S. & Stanley, P. M., *Computers in Australia – Usage and Effects*, Foundation for Australian Resources, Sydney, 1978.

Toynbee, Arnold, *Lectures on the Industrial Revolution of the 18th Century in England*, Longmans Green, London, 1884.

Thurow, Lester C., *The Zero-Sum Society*, Basic Books, New York, 1980.

Trevithick, J. A., *Inflation*, Penguin, London, 1977.

Tucker, K. A. (ed.), *Economics of the Australian Service Sector*, Croom Helm, London, 1977.

Valaskakis, Kimon, et al., *The Conserver Society: A Workable Alternative for the Future*, Harper & Row, New York, 1979.

Veblen, Thorstein, *The Theory of the Leisure Class* [1899], Allen & Unwin, London, 1924.

Vonnegut, Kurt Jr, *Player Piano*, Granada, London, 1977.

Walsh, Maximilian, *Poor Little Rich Country*, Penguin, Melbourne, 1979.

Ward, Barbara & Dubos, René, *Only One Earth*, Penguin, London, 1972.

Weber, Max, *The Protestant Ethic and the Spirit of Capitalism* [1905], Allen & Unwin, London, 1967.

Weizenbaum, Joseph, *Computer Power and Human Reason: From Judgement to Calculation*, W. H. Freeman, San Francisco, 1976.

Wiener, Norbert, *Cybernetics: or, Control and Communication in the Animal and the Machine*, Wiley, New York, 1948.

——, *The Human Use of Human Beings: Cybernetics and Human Beings*, Eyre & Spottiswoode, London, 1950.

Williams, Sir Bruce, et al., *Education, Training and Employment*, Report of Committee of Inquiry into Education and Training, 3 vols, Australian Government Publishing Service, Canberra, 1979.

Williams, Sir Edward, *Report of the Australian Commission of Inquiry into Drugs*, 5 vols, Australian Government Publishing Service, Canberra, 1980.

Williams, Raymond, *Keywords*, Fontana, London, 1976.

——, *The Long Revolution*, Penguin, London, 1961.

Williams, Shirley, 'Report on Unemployment', Report to OECD, in *Policy Studies*, London, April 1981.

Williams, Trevor I. (ed.), *A History of Technology*, vols VI and VII, Oxford University Press, Oxford, 1978.

Windschuttle, Keith, *Unemployment*, Penguin, Melbourne, 1979.

Winner, Langdon, *Autonomous Technology*, MIT Press, Cambridge, Mass., 1977.

Young, Michael, *The Rise of the Meritocracy*, Penguin, London, 1958.

——, & Willmott, Peter, *The Symmetrical Family*, Penguin, London, 1973.

Young, Mick, *I Want to Work*, Cassell, Sydney, 1979.

Zamyatin, Yevgeny I., *We* [1925], Penguin, London, 1972.

Zimmern, Sir Alfred, *The Greek Commonwealth*, Oxford University Press, Oxford, 1915.

Index